Annals of Mathematics Studies

Number 43

ANNALS OF MATHEMATICS STUDIES

Edited by Robert C. Gunning, John C. Moore, and Marston Morse

1. Algebraic Theory of Numbers, *by* HERMANN WEYL
3. Consistency of the Continuum Hypothesis, *by* KURT GÖDEL
6. The Calculi of Lambda-Conversion, *by* ALONZO CHURCH
10. Topics in Topology, *by* SOLOMON LEFSCHETZ
11. Introduction to Nonlinear Mechanics, *by* N. KRYLOFF *and* N. BOGOLIUBOFF
15. Topological Methods in the Theory of Functions of a Complex Variable, *by* MARSTON MORSE
16. Transcendental Numbers, *by* CARL LUDWIG SIEGEL
17. Problème Général de la Stabilité du Mouvement, *by* M. A. LIAPOUNOFF
19. Fourier Transforms, *by* S. BOCHNER *and* K. CHANDRASEKHARAN
20. Contributions to the Theory of Nonlinear Oscillations, Vol. I, *edited by* S. LEFSCHETZ
21. Functional Operators, Vol. I, *by* JOHN VON NEUMANN
22. Functional Operators, Vol. II, *by* JOHN VON NEUMANN
24. Contributions to the Theory of Games, Vol. I, *edited by* H. W. KUHN *and* A. W. TUCKER
25. Contributions to Fourier Analysis, *edited by* A. ZYGMUND, W. TRANSUE, M. MORSE, A. P. CALDERON, *and* S. BOCHNER
26. A Theory of Cross-Spaces, *by* ROBERT SCHATTEN
27. Isoperimetric Inequalities in Mathematical Physics, *by* G. POLYA *and* G. SZEGO
28. Contributions to the Theory of Games, Vol. II, *edited by* H. W. KUHN *and* A. W. TUCKER
29. Contributions to the Theory of Nonlinear Oscillations, Vol. II, *edited by* S. LEFSCHETZ
30. Contributions to the Theory of Riemann Surfaces, *edited by* L. AHLFORS *et al.*
31. Order-Preserving Maps and Integration Processes, *by* EDWARD J. MCSHANE
32. Curvature and Betti Numbers, *by* K. YANO *and* S. BOCHNER
33. Contributions to the Theory of Partial Differential Equations, *edited by* L. BERS, S. BOCHNER, *and* F. JOHN
34. Automata Studies, *edited by* C. E. SHANNON *and* J. MCCARTHY
35. Surface Area, *by* LAMBERTO CESARI
36. Contributions to the Theory of Nonlinear Oscillations, Vol. III, *edited by* S. LEFSCHETZ
37. Lectures on the Theory of Games, *by* HAROLD W. KUHN. In press
38. Linear Inequalities and Related Systems, *edited by* H. W. KUHN *and* A. W. TUCKER
39. Contributions to the Theory of Games, Vol. III, *edited by* M. DRESHER, A. W. TUCKER *and* P. WOLFE
40. Contributions to the Theory of Games, Vol. IV, *edited by* R. DUNCAN LUCE *and* A. W. TUCKER
41. Contributions to the Theory of Nonlinear Oscillations, Vol. IV, *edited by* S. LEFSCHETZ
42. Lectures on Fourier Integrals, *by* S. BOCHNER. In press
43. Ramification Theoretic Methods in Algebraic Geometry, *by* S. ABHYANKAR.

RAMIFICATION THEORETIC METHODS IN ALGEBRAIC GEOMETRY

by

Shreeram Abhyankar

PRINCETON, NEW JERSEY

PRINCETON UNIVERSITY PRESS

1959

L. C. Card 59-11071

Printed in the United States of America

PREFACE

These are the notes of a semester course that I gave at Columbia University in the winter session of 1955-56. There is little overlapping of the existing books on algebraic geometry with the contents of these notes. One novel aspect of this course is a simple proof of the local uniformization theorem on algebraic surfaces over algebraically closed ground fields of characteristic zero by a straightforward application of ramification theory (no continued fractions!). The contents of these notes is mainly based on the works of Abhyankar, Krull and Zariski referred to in the Bibliography. For an understanding of the main stream of discourse, Van der Waerden's Modern Algebra (Chapters I to VII and parts of XII, XIII, XIV) and Northcott's Ideal Theory (Chapters 1 to 4) are enough prerequisites. The writing down of these notes was partly supported by a National Science Foundation project in the Department of Mathematics at Harvard University, to which I am grateful.

The following is a novel aspect of the style of presentation. At several places proofs are omitted and are referred to the original sources, mainly Abhyankar and Zariski and at times to Krull. But this omission is done in well formed chunks and the statements of such results is made quite understandable so that even with these omissions the reader can travel a clearly visible pathway. Moreover partly because of these omissions, the supplying of many of which would have required a fair amount of preliminary material, this pathway is quick and short cutting across the subject carrying the reader a fair distance. It is well known that exhaustive treatments of algebraic geometry are liable to become rather formidable. The usual attempt of avoiding this formidability has mostly resulted in the production of a merely introductory exposition which leads the reader hardly to the gateway of the subject. Hence the author decided on the present method of exposition which in a semester would give the student a taste of at least some middle ground, no doubt very highly selective, of the subject. This can be achieved at the first reading of this monograph. While if during the second reading the student attempts to read up on the omissions, the general references for which are always given, that should serve as a good guide and motivation for reading of the original memoirs. Undoubtedly when he tries to study some of these memoirs he may have to go prepare some preliminaries, but that will always be indicated by the references in these papers and thus the student will not feel lost. This method of chronologically going backwards and forwards, I have always found instructive.

. The construction of this 'short' path owes its shortness only to a smaller extent to the above mentioned omissions. In the main it is due to things like the elegant, ramification theoretic, proof of uniformization; and the point of view that varieties are collections of local rings etc., i.e., to the prominence given to local rings. The first portion of the monograph can serve as a unified treatment of the elementary aspects of ramification theories required in algebraic geometry as well as in algebraic number theory. A possible sequel could include the following: (1) more useful lemmas concerning the topic of Section 7; (2) completions of local rings; (3) higher ramification groups of local rings (Hilbert-Krull); and (4) ramification theory of local rings for infinite extensions (yet to be developed).

Definitely there are many ways of approaching our subject and many angles to look at it from. This provides the reader with one.

Shreeram Abhyankar

Columbia University
Harvard University
October, 1956

CONTENTS

RAMIFICATION THEORETIC METHODS

IN

ALGEBRAIC GEOMETRY

INTRODUCTION

Let k be an algebraically closed field and let A_n be the affine n-space over k, i.e., A_n is the set of all n-tuples [called points of A_n] $(a) = (a_1, \ldots, a_n)$ with a_i in k. Let $f_i(X) = f_i(X_1, \ldots, X_n) = 0$ $(i = 1, \ldots, h)$, be a finite number of polynomial equations with coefficients in k. The set of points $(a) = (a_1, \ldots, a_n)$ of A_n which are common zeros of $f_1, \ldots, f_h$, i.e., for which $f_1(a) = f_2(a) = \ldots = f_h(a) = 0$ is called an <u>algebraic variety</u> V. $f_1 = 0, \ldots, f_h = 0$ is a set of defining equations of V. If f is the union of two proper subvarieties, then V is said to be <u>reducible</u>, otherwise V is irreducible. To an irreducible (algebraic) variety V can be attached an integer called the dimension of V. There are many ways of doing this. The intuitive description of dimension can be given in two ways: (1) It gives the number of degrees of freedom of a point allowed to move freely on V; (2) It gives the number of independent parameters in terms of which the n coordinates of a variable point of V can be expressed. There are many ways of making these intuitive concepts precise, e.g., (3) Set $d = -1$ if $V = \emptyset$. $d = 0$ if V consists of a single point, $d = 1$ if V consists of more than one point, and if any proper irreducible subvariety of V is either a point or the null set and so on, i.e., indutively having defined irreducible varieties of dimensions $-1, 0, 1, \ldots, s$, we define an irreducible variety V of dimension $s + 1$ by the

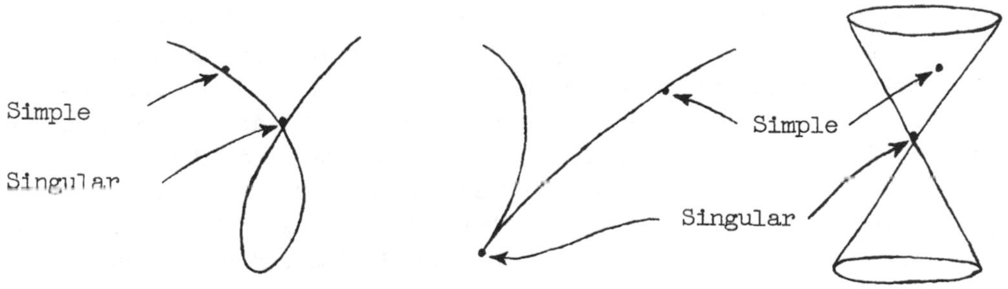

condition that V has irreducible proper subvarieties of dimension s and that any irreducible proper subvariety of V is of dimension at most s. If $d = 1, 2$ or 3 then V is respectively called a curve, a surface or a threefold. If V is a d-dimensional irreducible variety in A_n then $d \leq n$ and if $d = n$ then $V = A_n$. A point P of V is said to be a simple point of V if in the neighborhood of P, V has the structure of an affine A_d. If P is not simple, then P is singular. If V has no singular points either in A_n or at infinity then V is said to be nonsingular.

Let V and W be two irreducible varieties of dimension d. Let g be an algebraic correspondence (i.e., a correspondence defined by algebraic equations) between the points of V and W.

$$V \xrightarrow[g]{} W$$

It can be shown that the equations of g are rational if and only if g is single-valued at a generic point of V, i.e., single-valued at almost all points of V. Hence if g and g^{-1} are both rational, i.e., if g is birational, we get a correspondence between the points of V and W which is one-to-one almost everywhere. If such a birational correspondence between V and W exists, then V and W are said to be birationally equivalent. Projective equivalence is a very special case of birational equivalence (circle and ellipse). One of the central themes of algebraic geometry (which is the branch of mathematics studying algebraic varieties or equivalently "theory of equations in several variables") is to investigate the properties of an algebraic variety V which are invariant under birational transformations (e.g., genus of a curve V_1 = number of holes in V_1 if $k = C$ the field of complex numbers). Many questions of birational geometry can be studied only on a nonsingular variety. Hence, given a d-dimensional variety V it is of utmost importance to transform V birationally into a nonsingular variety or rather to know whether V can be thus transformed. This is the so-called problem of resolution of singularities. For curves, the solution is classical (Max Noether -- father of algebraic geometry -- Mathematische Annalen, 1876-1884) for $k = C$, and for arbitrary k was given around 1930). For surfaces, the (first complete) solution was given by Walker in 1935 (Annals) for $k = C$, by Zariski in 1939 [see Z3, Z5 in Bibliography] for fields of zero characteristic, and in 1954 see [A2 and A6] I completed the proof for fields of nonzero characteristic. In the case of zero characteristic, Zariski gave a solution for threefolds in 1944 [Annals]. The climax of this course will be a simplified solution of resolution of singularities for surface.

Let V be a d-dimensional irreducible variety, and let P be a point on V. Let K be the set of rational functions on V. Then K is a finitely generated extension of k of transcendental degree d and is called the <u>function field of</u> V. Let R be the set of rational functions on V which do not have a pole at P, then R is a noetherian ring with a unique maximal ideal M and is called the <u>local ring</u> of P on V. It turns out that a whole lot of information about V is concentrated in the function field K/k of V and the set of the local rings of the various points of V. Hence we shall adopt the attitude that a variety is defined by its local rings. To take care of finer aspects of V one has to consider branches (or, equivalently, sequences of points) on V; this will be taken care of by <u>valuations</u> (which represent the arithmetization of the idea of a branch) of K/k.

CHAPTER I: GENERAL RAMIFICATION THEORY

Summary. Let A be a domain integrally closed in its quotient field K, let K* be a finite separable algebraic extension of K and let A* be the integral closure of A in K*. Let [K* : K] = n. Let W and W* denote the sets of prime ideals in A and A*, respectively. Obviously $P^* \in W^* \Longrightarrow P = P^* \cap A \in W$, i.e., P* lies above the prime ideal P of A. We shall show that, conversely, for each P in W there is at least one member of W* lying above P, and there are at most n members of W* lying above P. In some sense, in general over $P \in W$ there lie n members (counted with their separable residue degrees) of W*, i.e., W* is an n-fold covering of W. Those $P \in W$ over which lie less than n "points" of W* are the "branch points" of this covering; they are given by a certain ideal in A called the discriminant ideal of A*/A. Now assume that K*/K is galois. Let $P^* \in W^*$ and $P = P^* \cap A \in W$. Then with the pair P*/P we associate two subgroups of the galois group of K*/K called the splitting group and the inertia group; their fixed fields are the splitting field and the inertia field, the splitting field is contained in the inertia field. As far as P* is concerned, in passing from K to the splitting field, P splits, acquires no residue field extension and stays unramified; in passing from the splitting field to the inertia field, P acquires the separable part of the residue field extension and stays unramified; and finally in passing from the inertia field to K*, there is pure ramification and the inseparable part of the residue field extension if any.

The above described situation of the n-fold covering W* of W is what we call "ramification theory" and it arises in two situations in algebraic geometry: (1) When A is the valuation ring of a valuation of K (in fact, for W we may take the set of all the prime ideals in all the various valuation rings of valuations of K. Then W is the Riemann manifold of K and W* is the Riemann manifold of K*); (2) When A is the affine coordinate ring of an algebraic variety V, in this case A is noetherian and W is in one-to-one correspondence with the set of irreducible subvarieties of V; A* is an affine coordinate ring of a derived normal model V^* of V in K^* and the map of W^* to W (whose inverse is n-valued) is the rational transformation from V^* to V. Now the local rings on algebraic varieties do not satisfy the valuation condition (except when they are one-dimensional) and the valuation rings do not satisfy the noetherian condition (unless they are real discrete). Hence, in this chaper, we develop a general ramification theory without assuming either of the two conditions; special aspects of the two cases will be dealt with in the following chapters. A breakdown of this chapter into various parts is clear from the contents.

1. Elementary lemmas from ideal theory. In this section we recall some lemmas from ideal theory; the proofs of most of them will be either left to the reader as exercises or referred to Northcott's *Ideal Theory* cited as N. By a ring we shall always means a commutative ring with identity. For the ease of reading, we shall try to use the letters: A, B for rings; H, I, J for ideals; P, Q for prime ideals; Q for primary ideals; R, S for semilocal rings and M, N for their maximal ideals.

Let A be a ring and let P, Q be ideals in A. Recall that P is a prime ideal if $P \neq A$ and one of the following three equivalent conditions is satisfied: (1) $a, b \in A$, $ab \in P$, $a \notin P$ implies $b \in P$; (2) H, I ideals in A, $HI \subset P$, $H \not\subset P$ implies $J \subset P$; (3) H, I are ideals in A which properly contain P implies HI is not contained in P. P is called a maximal ideal if $P \neq A$ and if H is an ideal in A with $P \subset H \neq A$ implies $P = H$. If P is prime and Q is primary, then Q is said to be primary for P if $a, b \in A$, $ab \in Q$, $a \notin Q$ implies $b \in P$. By $\sqrt{Q}$ we denote the radical of Q. If $P + Q = A$ then P and Q are said to be coprime.

LEMMA 1.1. Suppose $Q \subset P \subset \sqrt{Q}$ and that $ab \in Q$, $a \notin p \Longrightarrow b \in Q$. Then either $Q = P = A$ or P is prime and Q is primary for P. [N, Lemma 1, p. 11.]

LEMMA 1.2. Let P_1, P_2 be prime ideals in A which are coprime and let Q_1, Q_2 be ideals in A which are primary for P_1, P_2, respectively. Then Q_1, Q_2 are coprime.

PROOF. Otherwise, there would exist a maximal ideal M in A containing $Q_1 + Q_2$. $Q_i \subset M$ and M is prime implies $P_i \subset M$ for $i = 1, 2$, so that $P_1 + P_2 \subset M \neq A$ which is a contradiction.

LEMMA 1.3. Let $H_1, \ldots, H_n$ be pairwise coprime ideals in A. Then

$$\prod_{i=1}^{n} H_i = \bigcap_{i=1}^{n} H_i .$$

Let $a_1, \ldots, a_n$ be given elements in A. Then there exists a in A such that $a \equiv a_i \bmod H_i$ for $i = 1, \ldots, n$.

PROOF. For $i \neq j$, H_i and H_j are coprime implies that there exists $h_{ij} \in H_i$ and $h_{ji} \in H_j$ such that $h_{ij} + h_{ji} = 1$ so that $h_{ji} \equiv 1 \bmod H_i$ and $0 \bmod H_j$. Let

$$h_i = \prod_{j \neq i} h_{ji} .$$

Then $h_i \equiv 1 \bmod H_i$ and $0 \bmod H_j$ for all $j \neq i$. Let $a = a_1 h_1 + \ldots + a_n h_n$. Then $a \equiv a_i \bmod H_i$ for $i = 1, \ldots, n$. To prove $\Pi H_i = \cap H_i$, we make induction on n; obvious for $n = 1$, so let $n > 1$ and assume true for $n - 1$. Then

$$\bigcap_{i=1}^{n} H_i \subset \bigcap_{i\neq 1} H_i = \prod_{i \neq 1} H_i$$

and

$$\bigcap_{i=1}^{n} H_i \subset \bigcap_{i\neq 2} H_i = \prod_{i \neq 2} H_i \quad ;$$

since $H_1 + H_2 = A$ we have

$$\bigcap_{i=1}^{n} H_i = (H_1 + H_2)\left(\bigcap_{i=1}^{n} H_i\right) = H_1\left(\bigcap_{i=1}^{n} H_i\right) + H_2\left(\bigcap_{i=1}^{n} H_i\right)$$

$$\subset H_1\left(\bigcap_{i\neq 1}^{n} H_i\right) + H_2\left(\bigcap_{i\neq 2}^{n} H_i\right) = H_1\left(\prod_{i\neq 1}^{n} H_i\right) + H_2\left(\prod_{i\neq 2}^{n} H_i\right) = \prod_{i=1}^{n} H_i \ ,$$

i.e., $\cap H_i = \Pi H_i$.

LEMMA 1.4. Let $P_1, \ldots, P_n$ be prime ideals in A and Q an ideal in A such that $Q \not\subset P_i$ for $i = 1, \ldots, n$. Then there exists $a \in A$ such that $a \in Q$ and $a \notin P_i$ for $i = 1, \ldots, n$.

PROOF. A proof by induction on n is given in Proposition 6, p. 12 of [N]; a direct proof is as follows: omitting each P_j which is contained in some P_i with $i \neq j$ we may assume that $P_j \not\subset P_i$ whenever $j \neq i$; then there exists $a_{ij} \in P_j, \notin P_i$; fix $b_i \in Q$ with $b_i \notin P_i$. Let

$$a_i = b_i \prod_{j \neq i} a_{ij} \quad .$$

Then $a_i \in Q$, $a_i \notin P_i$ and $a_i \in P_j$ if $j \neq i$. Let $a = a_1 + \cdots + a_n$. Then $a \in Q$ and $a \notin P_i$ for $i = 1, \ldots, n$.

LEMMA 1.5 If the set M of all the nonunits of A is an ideal then M is the unique maximal ideal in A. Conversely, if A contains a unique maximal ideal M then M is the set of all the nonunits of A.

If a ring R contains only a finite number of maximal ideals $M_1, \ldots, M_n$ then R is called a semilocal ring; we shall express this by saying that $(R; M_1, \ldots, M_n)$ is a semilocal ring. If $n = 1$ we shall also write (R, M_1) for $(R; M_1)$.

LEMMA 1.6. Let B be a ring and A a subring of B. Then we have:

(1) For an ideal I in A, $I \subset IB \cap A$ always, and $I = IB \cap A$ if and only if there exists an ideal J in B with $J \cap A = I$, i.e., if and only if I is a contracted ideal for the ring extension B/A.

(2) For an ideal J in B, $J \supset (J \cap A)B$ always, and $J = (J \cap A)B$ if and only if there exists an ideal I in B with $IB = J$, i.e., if and only if J is an extended ideal for the ring extension B/A.

(3) $\left.\begin{array}{l} I \longrightarrow IB = J \\ J \cap A \longleftarrow J \end{array}\right\}$ is a one-to-one correspondence between the contracted and the extended ideals, and it preserves the ideal theoretic operations, $+$, $\cap$, $\cdot$, $:$.

Now let S be a nonempty multiplicative subset of nonzero divisors in A; by A_S we denote the quotient ring of A with respect to S, i.e., the set of elements of the total quotient ring of A which are of the form a/b with a in A and b in S. If P is a prime ideal in A, we write A_P instead of A_{A-P}.

LEMMA 1.7. Let H be an ideal in A. Then HA_S coincides with the set of elements a/b with a in H and b in S. [N, Lemma 3, p. 42].

LEMMA 1.8. Let P be a prime ideal in a ring A. Then PA_P is the unique maximal ideal in A_P.

PROOF. This follows from the next lemma; a direct proof is as follows: Let $x \in A_P$, $x \notin PA_P$. $x \in A_P$ implies $x = a/b$ with $a, b \in A$, $b \notin P$. Since $x \notin PA_P$, $a \notin P$ by Lemma 1.7 and hence $x^{-1} = b/a \in A_P$, i.e., PA_P contains all the nonunits in A_P. Since any element in PA_P is of the form a/b with $a \in P$, $b \notin P$ so that $a \neq b$; therefore PA_P does not contain 1 and hence it does not contain any unit of A_P. Hence, by Lemma 1.5, PA_P is the unique maximal ideal in A_P.

LEMMA 1.9. Let P be a prime ideal in ring A, let $A^* = A_P$ and $P^* = PA^*$. Then:

(1) Every ideal in A^* is extended.

(2) If H is an ideal in A then $HA^* = A^*$ if and only if $H \cap P \neq H$.

(3) Contraction and extension sets up a one-to-one correspondence between all the prime ideals of A^* and those prime ideals of A which are contained in P. Let P_1^* and P_1 be two such corresponding prime ideals in A^* and A; then we also have a one-to-one correspondence between ideals in A^* which are primary for P_1^* and ideals in A which are primary for P_1.

(2') If Q is an ideal in A which is primary for a prime ideal not contained in P then $QA^* = A^*$, i.e., Q is lost in A^*.

(4) If $H_1, \ldots, H_n$ are ideals in A then $(H_1 \cap \ldots \cap H_n)A^* = H_1A^* \cap \ldots \cap H_nA^*$.

(5) A/P and A^*/P^* are canonically isomorphic.

(6) If A is noetherian, then so is A^*.

PROOF. Most parts are in [N, Propositions 9 to 11, pp. 41-43].

LEMMA 1.10. Let W be the set of all the maximal ideals in a ring A. Then $A = \cap_{P \in W} A_P$.

Now let A be a ring and let $A_1, \ldots, A_n$ be nonzero ideals in A such that $A = A_1 \oplus \cdots \oplus A_n$ (direct sum). Then A_i is a ring and this is a direct product; let e_i be the identity in A_i, then $A_i = e_iA$ and $e_1, \ldots, e_n$ are orthogonal idempotents (i.e., $e_i^2 = e_i$ and $e_ie_j = 0$ if $i \neq j$) with $1 = e_1 + \ldots + e_n$.

PROOF. For $a_i \in A_i$, $a_i \notin A_j$ with $i \neq j$ we have $a_ia_j \in A_i \cap A_j = \{0\}$, i.e., $a_ia_j = 0$; therefore the decomposition $A = A_1 \oplus \cdots \oplus A_n$ is a direct product. Let e_i be the unique element of A_i such that $1 = e_1 + \ldots + e_n$. Then for $i \neq j$, $e_ie_j = 0$ and hence $1 = 1^2 = e_1^2 + \ldots + e_n^2$, i.e., $e_i^2 = e_i$; for $a_i \in A_i$, $a_i = a_i1 = a_ie_1 + a_ie_2 + \ldots + a_ie_n = a_ie_i$ so that e_i is the identity in A_i and $A_i = e_iA$.

Conversely, let $e_1, \ldots, e_n$ be orthogonal idempotents in A with $1 = e_1 + \ldots + e_n$ and let $A_i = e_iA$. Then it is easily verified that we have the direct sum $A = A_1 \oplus \cdots \oplus A_n$. One instance in which such a decomposition of the identity into orthogonal idempotents occurs is the following: Suppose we have $\{0\} = H_1H_2 \cdots H_n$ where the H_i are pairwise coprime ideals in A. Since the H_i are pairwise coprime there exists a_{ij} in A such that $a_{ij} \equiv 0 \bmod H_j$ if $j \neq i$ and $a_{ij} \equiv 1 \bmod H_i$. Let

$$e_i = \prod_{j \neq i} a_{ij} .$$

Since by Lemma 1.3, $\{0\} = H_1 H_2 \cdots H_n = H_1 \cap H_2 \cap \cdots \cap H_n$, it it follows that $e_i^2 = e_i$, $e_i e_j = 0$ if $i \neq j$ and $1 = e_1 + \cdots + e_n$. Then we have the direct sum decomposition $A = A_1 \oplus \cdots \oplus A_n$ where $A_i = e_i A$. Let now f_i be the canonical homomorphism: $A \longrightarrow A/H_i = A_i^*$ and let f be the homomorphism of A into $A_1^* \oplus \cdots \oplus A_n^*$ given by: $f(a) = (f_1(a), \ldots, f_n(a))$. Then f_i maps A_i isomorphically onto A_i^* and for $j \neq i$ it is zero on A_j. Therefore f maps A isomorphically onto $A_1^* \oplus \cdots \oplus A_n^*$ and hence we may identify A with $A_1^* + \cdots + A_n^*$.

2. <u>Primary decomposition in nonnoetherian rings</u>. The aim of this section is to prove the following proposition which gives a primary decomposition for certain types of ideals in an arbitrary (not necessarily noetherian) ring A.

PROPOSITION 1.11. Let H be an ideal in A with $H \neq A$ such that there are only a finite number of prime ideals $P_1, \ldots, P_n$ containing H and each of the P_i is a maximal ideal. Then $H = Q_1 \cap \cdots \cap Q_n$ $(= Q_1 Q_2 \cdots Q_n)$ where Q_i is primary for P_i. Furthermore the Q_i are unique, i.e., if $A = Q_1^* \cap \cdots \cap Q_n^*$ where Q_i^* is primary for P_i, then $Q_i^* = Q_i$ for $i = 1, \ldots, n$.

The proof of this proposition will be preceded by several lemmas. Let H and I be ideals in A and h an element of A. H is said to be prime to I if $(I : H) = I$; h is said to be prime to I if hA is prime to I, i.e., if $ah \in I$ with $a \in A$ implies $a \in I$. h is said to be a zero divisor mod I if for some a in A with $a \notin I$ we have $ah \in I$, i.e., if the residue class of h modulo I is a zero divisor in A/I; H is said to be a zero divisor mod I if each element of H is a zero divisor mod I. Note that h is a zero divisor mod I if and only if hA is a zero divisor mod I. Also observe that h is prime to I if and only if h is not a zero divisor mod I. Now assume $I \neq A$ and let $S(I)$ denote the set of elements a in A which are prime to I; then $S(I)$ is a nonempty multiplicative subset of A and H (respectively h) is a zero divisor mod I if and only if $H \cap S(A) = \emptyset$ (respectively $h \notin S(I)$). If S is any nonempty multiplicative set in A then by I_S we shall denote the set of elements a in A such that $as \in I$ for some $s \in S$; it is clear that I_S is an ideal in A with $I \subset I_S \subset A$; I_S is called an isolated component of I and also the S-component of I; it follows from Lemma 1.7 that if S contains no zero divisors of A then $I_S = A \cap IA_S$. If P is a prime ideal in A we shall write I_P for I_{A-P} and call it the P-component of I.

LEMMA 1.12. Let S be a nonempty multiplicative subset of A which is disjoint from a given ideal $I \neq A$ (in particular we may take $I = \{0\}$). Let W be the set of all ideals in A which contain I and are disjoint from S. Then W contains maximal elements and they are necessarily prime.

Furthermore, let I_1 be an ideal contained in I and let W_1 be the set of all ideals in A which contain I_1 and are disjoint from S. Then any maximal element of W_1 is also a maximal element of W.

PROOF. Obviously W has the Zorn property. Let P be a maximal element of W. Let H_1, H_2 be ideals in A which properly contain P; then for $i = 1, 2$ there exists $h_i \in H_i$ with $h_i \in S$ so that $h_1 h_2 \in H_1 H_2$ and $h_1 h_2 \in S$ so that $h_1 h_2 \notin P$; hence $H_1 H_2 \not\subset P$. Therefore P is prime. The last assertion is obvious.

In particular, if $S = S(I)$, we shall say that P is a maximal zero divisor ideal of I and that I_P is a principal component of I.

If P is a prime ideal in A such that $I \subset P$ and there is no other prime ideal between I and P then we shall say that P is a minimal prime ideal of I.

LEMMA 1.13. Let H and I be ideals in A different from A such that H is a zero divisor mod I. Then there exists a maximal zero divisor P mod I such that $P \supset H$.

PROOF. H is a zero divisor mod I implies that $H + I$ is a zero divisor mod I and hence $H + I \cap S(I) = \emptyset$. By Lemma 1.12, there exists a maximal zero divisor P mod I such that $P \supset H + I \supset H$.

LEMMA 1.14. Let I be an ideal in A different from A. Then I = the intersection of all its principal components.

PROOF. Clearly I is contained in the intersection. Now let u be an element in the intersection and let $H = (I : uA)$. Then [P is a maximal zero divisor ideal of I] $\Longrightarrow$ [$u \in I_P$] $\Longrightarrow$ [there exists v in A, $v \notin P$ such that $uv \in I$ so that $v \in H$] $\Longrightarrow$ [$H \not\subset P$]. Hence, by Lemma 1.13, H is a nonzero divisor ideal mod I, i.e., H contains a nonzero divisor h mod I. Now $h \in H$ implies $hu \in I$ and hence $u \in I$.

LEMMA 1.15. Let I be an ideal in A different from A and let P be a minimal prime ideal of I. Then (1) I_P is primary for P and (2) P is a zero divisor mod I.

PROOF. To prove (1) we have to show that (i) $I_P \subset P \subset \sqrt{I_P}$, and that (ii) $u, v \in A$ with $uv \in I_P$ and $u \notin P$ implies $v \in I_P$. Now $u \in I_P$ implies there exists $v \in A$, $v \notin P$ such that $uv \in I$ so that $uv \in P$ and hence $u \in P$. Therefore $I_P \subset P$.

Next, given $u \in P$ let S be the smallest multiplicative set containing u and $A - P$. If I were disjoint from S then by Lemma 1.12 there would exist a prime ideal Q containing I and disjoint from S. But Q is disjoint from S implies that Q is disjoint from $A - P$, i.e., $Q \subset P$ which in view of the minimality of P implies that $Q = P$ so that

$u \notin P$ which is a contradiction. Therefore I is not disjoint from S, i.e., there exists $v \in A - P$ and a nonnegative integer n such that $u^n v \in I$. Now $u^n v \in I$ and $v \notin P$ imply that $u^n \in I_P$, i.e., $u \in \sqrt{I_P}$. Therefore $P \subset \sqrt{I_P}$. This proves (i).

Next, let $u, v \in A$ with $uv \in I_P$ and $u \notin P$. Since $uv \in I_P$, there exists $w \in A$, $w \notin P$ such that $w(uv)$ is in I. Since $w \notin P$ and $u \notin P$ we have $wu \notin P$; since $(wu)v = w(uv)$ is in I we conclude that $v \in I_P$. This proves (ii) and hence (1).

Finally, let $u \in P$. Then, in view of (i), $u^n \in I_P$ for some n whence $u^n v \in I$ for some v in A with $v \notin P$ so that $v \notin I$. Therefore u^n is a zero divisor mod I and hence u is also a zero divisor mod I. Thus P is a zero divisor mod I which proves (2).

PROOF OF PROPOSITION 1.11. By 1.15 (2); $P_1, \ldots, P_n$ are exactly the maximal zero divisor ideals of H. Let $Q_i = I_{P_i}$. By 1.15 (1), Q_i is primary for P_i and by 1.14,

$$H = \bigcap_{i=1}^{n} Q_i \quad .$$

Since the P_i are pairwise coprime, by 1.2 and 1.3 we have $\bigcap Q_i = \prod Q_i$. The uniqueness follows from [N], Theorem 2 on p. 15.

3. Integral dependence. Let A be a ring and B an overring of A (with the same identity). Let P and Q be prime ideals in A and B respectively; if $Q \cap A = P$ then we shall say that Q lies above P or P lies below Q. An element u of B said to be integral over A if there exist $a_0, a_1, \ldots, a_{n-1}$ in A such that $u^n + a_{n-1}u^{n-1} + \ldots + a_0 = 0$. If every element of B is integral over A we shall say that B is integral over A and express this by saying that B/A is integral. If A is a domain such that A contains every element of its quotient field which is integral over A, then A will be said to be normal.

LEMMA 1.16 (Kronecker). Let A be a normal domain with quotient field K. Let $f(X)$, $g(X)$ be monic polynomials in $K[X]$ and $h(X) = f(X)g(X)$. Then $h(X) \in A[X]$ implies $f(X), g(X) \in A[X]$.

Let K^* be an overfield of K. Then $u \in K^*$ with u/A integral implies that the minimal monic polynomial of u/K is in $A[X]$.

PROOF. The second part follows from the first. To prove the first part, let L be a root field of $h(X)$ over K. Let

$$f(X) = \sum_{i=0}^{n} f_{n-i}X^i = \prod_{i=1}^{n} (X - u_i)$$

with $f_i \in K$, $f_o = 1$ and $u_i \in L$. Let $p_i(Y_1, \ldots, Y_n)$ be $(-1)^i$ times the i-th elementary symmetric function in $Y_1, \ldots, Y_n$, so that $f_i = p_i(u_1, \ldots, u_n)$. Now $f(u_i) = 0$ implies $h(u_i) = 0$ which implies that u_i/A is integral. Since the integral closure of A in L is a domain, $u_1, \ldots, u_n$ are integral over A implies that $f_i = p_i(u_1, \ldots, u_n)$ is integral over A. Since A is normal and $f_i \in K$ we must have $f_i \in A$ for $i = 1, \ldots, n$, i.e., $f(X) \in A[X]$. Similarly, $g(X) \in A[X]$.

Another proof of the above lemma using valuations will be given in Chapter II, see Corollary 2.27.

LEMMA 1.17. Let A be a normal domain and P a prime ideal in A. Then A_P is normal.

PROOF. Let u be an element of the quotient field of A which is integral over A_P. Then

$$u^n + a_1 u^{n-1} + \ldots + a_n = 0 \quad , \qquad \text{with } a_i \in A_P \ .$$

$a_i \in A_P$ implies that $a_i = b_i/c_i$ with $b_i, c_i \in A$ and $c_i \notin P$. Let

$$e = \prod_{i=1}^{n} c_i \qquad \text{and} \qquad d_i = b_i \prod_{j \neq i} c_j \ .$$

Then $a_i = d_i/e$ with $d_i, e \in A$ and $e \notin P$. Let $v = eu$ and $a_i' = d_i e^{i-1}$. Then

$$v^n + a_1' v^{n-1} + \ldots + a_n' = e^n u^n + d_1 e^{n-1} u^{n-1} + d_2 e e^{n-2} u^{n-2}$$
$$+ \ldots + d_n e^{n-1} = e^n\Big[u^n + (d_1/e)u^{n-1} + (d_2/e)u^{n-2}$$
$$+ \ldots + (d_n/e)\Big] = 0 \quad ,$$

and $a_i' \in A$. Since A is normal, $v \in A$. Since $e \notin P$, $u = (v/e) \in A_P$.

LEMMA 1.18. Let B be a domain and A a subdomain of B such that B/A is integral. Then A is a field if and only if B is a field.

PROOF. Assume A is a field and let $0 \neq u \in B$. Then u/A is integral and hence u/A is algebraic so that $u^{-1} \in A[u] \subset B$. Therefore B is a field.

Now assume that B is a field and let $0 \neq u \in A$. Then $u^{-1} \in B$ and hence u^{-1}/A is integral, i.e., $(u^{-1})^n + a_{n-1}(u^{-1})^{n-1} + \ldots + a_o = 0$ with $a_i \in A$ so that $u^{-1} = -(a_{n-1} + a_{n-2}u + \ldots + a_o u^{n-1}) \in A$. Therefore A is a field.

LEMMA 1.19. Let B be a ring and A a subring of B such that B/A is integral. Let P be a prime ideal in A lying below a prime ideal Q in B. Then P is maximal if and only if Q is maximal.

PROOF. We may canonically assume that $B/Q \supset A/P$. Then $(B/Q)/(A/P)$ is integral. Now apply Lemma 1.18.

LEMMA 1.20. Let B be a ring and A a subring of B such that B/A is integral and let P be a prime ideal in A. Then there exists a prime ideal Q in B lying above P.

PROOF. Let W be the set of ideals J in B for which $J \cap A \subset P$. Then W has the Zorn property and hence has a maximal element Q. We assert that Q is prime. For $u, v \in B$; $u, v \notin Q$ and $uv \in Q \Longrightarrow (Q + uB) \cap A \not\subset P$, $(Q + vB) \cap A \not\subset P \Longrightarrow$ there exist $u^*, v^* \in A$, $\notin P$ such that $u^* \equiv au \pmod{Q}$ and $v^* \equiv bv \pmod{Q}$ with $a, b \in B \Longrightarrow u^*v^* \equiv abuv \pmod{Q} \Longrightarrow u^*v^* \in P$ which is a contradiction. Hence Q is prime. Now assume if possible that $Q \cap A \neq P$. Then there exists $u \in P$, $\notin Q$ so that $Q + uB$ properly contains Q and hence $(Q + uB) \cap A \not\subset P$, i.e., there exists $v \in A$, $\notin P$ and $b \in B$ such that $v \equiv bu \pmod{Q}$. Since B/A is integral, we have

$$b^n + a_{n-1}b^{n-1} + \dots + a_o, \qquad \text{with } a_i \in A \ .$$

Therefore

$$(bu)^n + ua_{n-1}(bu)^{n-1} + u^2a_{n-2}(bu)^{n-2} + \dots + u^na_o = 0 \ .$$

Since $v \equiv bu \pmod{Q}$ we have

$$v^n + (ua_{n-1})v^{n-1} + (u^2a_{n-2})v^{n-2} + \dots + (u^na_o) \equiv 0 \pmod{Q} \ .$$

Since the left-hand side is in A, it must be $\equiv 0 \pmod{P}$. Since $u \in P$ we have $v^n \in P$ and hence $v \in P$ which is a contradiction. Therefore $Q \cap A = P$.

LEMMA 1.21. Let B be a ring and A a subring of B, and let P be a prime ideal in A. Then there exists a prime ideal Q in B lying above $P \Longleftrightarrow PB \cap A = P$.

PROOF. ($\Longrightarrow$) Obvious. ($\Longleftarrow$) Let W be the set of ideals J in B such that $J \cap A = P$. Then W has the Zorn property and hence a maximal element Q. As in the proof of 1.20, it follows that Q is prime.

LEMMA 1.22. Let B be a ring and A a subring of B such that B/A is integral. Let P and P^* be prime ideals in A such that

$P \supset P^*$ and let Q^* be a prime ideal in B lying above P^*. Then there exists a prime ideal Q in B lying above P such that $Q \supset Q^*$. [Going up theorem for B/A.]

PROOF. Apply 1.20 to $(B/Q^*)/(A/P^*)$.

LEMMA 1.23. Let B be a domain and A a subdomain of B such that B/A is integral and let Q be a nonzero ideal in B. Then $Q \cap A \neq \{0\}$.

PROOF. Fix $0 \neq u$ in Q and let $u^n + a_{n-1}u^{n-1} + \dots + a_o = 0$ with $a_i \in A$ be an equation of integral dependence of u/A of least possible degree. Then $0 \neq a_o \in Q \cap A$.

LEMMA 1.24. Let B be a ring, and A a subring of B such that B/A is integral; let Q be a prime ideal in B lying above the prime ideal P in A, and let J be an ideal in B properly containing Q. Then $J \cap A$ properly contains P.

PROOF. Apply 1.23 to $(B/Q)/(A/P)$.

LEMMA 1.24 A. Let B be a domain and A a subdomain such that B/A is integral and let H be an ideal in A. Then $\sqrt{HB} = \{b \in B \mid b^m + h_1b^{m-1} + \dots + h_m = 0,$ with $h_i \in H\}$.

PROOF. Let $b^m + h_1b^{m-1} + \dots + h_m = 0$ with $b \in B$ and $h_i \in H$. Then $b^m \in HB$ and hence $b \in \sqrt{HB}$. Conversely, let $d \in \sqrt{HB}$, then $d^k \in HB$ for some k; if d^k would satisfy an equation of the above form, then so would d, hence we may assume $d \in HB$. If $d = hb$ with $h \in H$ and $b \in B$ and if $b^m + a_1b^{m-1} + \dots + a_m = 0$ is an equation of integral dependence of b/A then $d^m + h_1d^{m-1} + \dots + h_m = 0$ with $h_i = h^ia_i \in H$ for $i = 1, \dots, m$. Since every element of HB is a finite sum of elements of type hb, it is enough to show that if elements d and e of B satisfy equations of the above form, then so does $d + e$. Let

$$d^m + h_1d^{m-1} + \dots + h_m = 0 , \qquad \text{with } h_i \in H ,$$

and

$$e^n + k_1e^{n-1} + \dots + e_n = 0 , \qquad \text{with } k_i \in H .$$

Let the products d^ie^j, $0 \leqq i \leqq m - 1$, $0 \leqq j \leqq n - 1$, be designated as $f_1, f_2, \dots, f_{mn}$. Then every power d^r with $r \geqq m$ can be written as a linear combination of $1, d, \dots, d^{m-1}$ with coefficients in H and every power e^s with $s \geqq n$ can be written as a linear combination of $1, e, \dots, e^{n-1}$ with coefficients in H. Therefore every product d^re^s with $r + s \geqq m + n - 1$ can be written as a linear combination of $f_1, \dots, f_{mn}$ with coefficients in H. Hence for $p = m + n - 1$ we have

$$(d + e)^p f_i = \sum_{j=1}^{mn} h_{ij} f_j , \qquad \text{with } h_{ij} \in H \text{ and } i = 1, \ldots, mn.$$

Hence $f_i \det((d + e)^p \delta_{ij} - h_{ij}) = 0$, for $i = 1, \ldots, mn$. Since one of the f_i is 1, the determinant is zero and this is an equation for $d + e$ of the required form.

PROPOSITION 1.24B. Let A be a normal domain and B an overdomain of A which is integral over A. Let $P \subset Q$ be prime ideals in A and let Q^* be a prime ideal in B lying over Q. Then there exists a prime ideal P^* in B lying over P with $P^* \subset Q^*$. [Going down theorem for B/A.]

PROOF. Let S be the multiplicative subset of B of elements of the form ab with $a \in A - P$ and $b \in B - Q^*$. Suppose if possible that there exist $a \in A - P$ and $b \in B - Q^*$ with $ab \in PB$. By 1.24A, there exists

$$h(x) = x^n + h_{n-1}x^{n-1} + \ldots + h_o , \qquad \text{with } h_i \in P ,$$

such that $h(ab) = 0$. Let

$$f(x) = x^m + f_{m-1}x^{m-1} + \ldots + f_o$$

be the monic minimal polynomial of ab over the quotient field of A and let

$$g(x) = h(x)/f(x) = x^q + g_{q-1}x^{q-1} + \ldots + g_o .$$

By 1.16, $f(x), g(x) \in A[x]$. Let $h_n = f_m = g_q = 1$. Let i be minimum such that $f_i \notin P$ and let j be minimum such that $g_j \notin P$. Then

$$h_{i+j} = f_i g_j + \text{terms in } P ,$$

and hence $f_{i+j} \notin P$ so that $i + j = n$, i.e., $i = m$ and $j = q$. Hence $f_t \in P$ for $t = 0, 1, \ldots, m - 1$. Now

$$f^*(x) = x^m + f^*_{m-1}x^{m-1} + \ldots + f^*_o, \qquad \text{with } f^*_i = f_i/a^{m-i}$$

is the minimal monic polynomial of b over the quotient field of A. By 1.16, $f^*_i \in A$. Since $f_i = a^{m-i}f^*_i$, $a \notin P$ and $f_i \in P$, we must have $f^*_i \in P$. Therefore

$$b^m = -(f^*_{m-1}b^{m-1} + \dots + f^*_o) \in PB \subset Q^* \quad ,$$

which is a contradiction. Therefore $S \cap PB = \emptyset$. Hencey by 1.12, there exists a prime ideal P^* in B such that $P \cap S = \emptyset$ and $P^* \supset PB$. Then $P^* \cap S = \emptyset \implies P^* \cap (B - Q^*) = \emptyset \implies P^* \subset Q^*$. Also $P^* \cap S = \emptyset \implies (P^* \cap A) \cap (A - P) = \emptyset \implies P^* \cap A \subset P$. Since $P^* \cap A \supset PB \cap A \supset P$, we have: $P^* \cap A = P$.

PROPOSITION 1.25. Let A be a normal domain with quotient field K and let A^* be the integral closure of A in a finite normal extension K^* of K with $[K^* : K] = n$. Let P be a prime ideal in A. Then there are only a finite number of prime ideals $P^*_1, \dots, P^*_m$ in A^* lying above P, with $m \leqq n$; and $P^*_1, \dots, P^*_m$ form a complete set of K-conjugates.

PROOF. By 1.20, there exists a prime ideal P^*_1 in A^* lying above P. Let $P^*_2, \dots, P^*_m$ be the other K-conjugates of P^*_1; we must then have $m \leqq n$ and $P^*_i \cap A = P$ for $i = 1, \dots, m$. Suppose if possible that Q^* is a prime ideal in A^* lying above P with $Q^* \neq P^*_i$ for $i = 1, \dots, m$. By 1.24, $Q^* \not\subset P^*_i$ for $i = 1, \dots, m$, and hence by 1.4 there exists $u \in Q^*$ such that $u \notin P^*_i$ for $i = 1, \dots, m$. Then none of the K-conjugates of u can be in any P^*_i and hence $N_{K^*/K}(u) \notin P^*_i$ for $i = 1, \dots, m$. But $N_{K^*/K}(u) \in Q^* \cap A \subset P^*_1$ which is a contradiction.

LEMMA 1.26. Let A be a normal domain with quotient field K, let K^* be a finite algebraic extension of K, let A^* be a domain such that $A \subset A^* \subset K^*$ and such that A^*/A is integral, and let P be a prime ideal in A. Then there exist only a finite number of prime ideals in A^* lying above P.

PROOF. Pass to the integral closure of A in a finite normal extension of K containing K^* and invoke 1.25.

DEFINITION 1.27. Let (R, M) be a normal local domain with quotient field K and let R^* be the integral closure of R in an extension field K^* of K. Let $M^*_1, M^*_2, \dots,$ be the prime ideals in R^* lying above M. It follows by 1.19 that $M^*_1, M^*_2, \dots,$ are exactly the maximal ideals in R^*. Let $S_i = R^*_{M^*_i}$ and $N_i = M^*_i S_i$. We shall say that $(S_1, N_1), (S_2, N_2), \dots,$ are the local rings in K^* lying above (R, M). If $[K^* : K]$ is finite it follows by 1.26 that there are only a finite number of local rings in K^* lying above (R, M); let $M^*_1, \dots, M^*_n$ be the maximal ideals in R^*: then we shall use the expression: Let [the semilocal ring] $(R^*; M^*_1, \dots, M^*_n)$ be the integral closure of R [or of (R, M)] in K^*; if furthermore K^*/K is normal, then it follows by 1.25 that $(S_1, N_1), \dots, (S_n, N_n)$ form a complete K-conjugate set.

LEMMA 1.28. Let A be a domain with quotient field K and let P be a prime ideal in A. Let $R = A_P$ and $M = PR$. Let K^* be a field extension of K, let R^* be the integral closure of R in K^* and let A^* be a subdomain of K^* integral over A. Then it is clear that $R^* \supset A^*$. Let $M_1^*, M_2^*, \ldots,$ be the prime ideals in R^* lying above M and let $P_i^* = M_i^* \cap A^*$. Then (1) $P_1^*, P_2^*, \ldots,$ are exactly the prime ideals in A^* lying above P (although they need not all be distinct). Now assume that A^* is the integral closure of A in K^*. Then (2) $P_1^*, P_2^*, \ldots,$ are all distinct and $R^*_{M_i^*} = A^*_{P_i^*}$ for $i = 1, 2, \ldots$.

PROOF. (1) Obviously $P_1^*, P_2^*, \ldots,$ lie above P. Now let P^* be any prime ideal in A^* lying above P. Let $S = A^*_{P^*}$, $N = P^*S$ and S^* = the integral closure of S in K^*. Let N^* be a maximal ideal in S^*. Then $N^* \cap S = N$. Now

$$A^* \supset A \qquad \text{and} \qquad P^* \cap A = P \quad .$$

Therefore

$$S \supset R \qquad \text{and} \qquad N \cap R = M \quad .$$

Therefore

$$S^* \supset R^* \qquad \text{and} \qquad N^* \cap R^* = M_i^* \qquad \text{for some } i \quad .$$

Therefore

$$P^* = N^* \cap A^* = M_i^* \cap A^* = P_i^* \quad .$$

(2) Obviously $R^*_{M_i} \supset A^*_{P_i^*}$. Also $A^*_{P_i^*} \supset R$ and by 1.17, $A^*_{P_i^*}$ is normal. Therefore

$$A^*_{P_i^*} \supset R^* \quad \text{and} \quad P_i^* A^*_{P_i^*} \cap R^* = M_j^* \qquad \text{for some } j \quad .$$

Therefore

$$R^*_{M_j^*} \subset A^*_{P_i^*} \subset R^*_{M_i^*} \quad .$$

Therefore

$$M_j^* = R^* \cap M_j^* R^*_{M_j^*} = R^* \cap P_i^* A^*_{P_i^*} = R^* \cap R^*_{M_i^*} = M_i^* \quad .$$

Therefore $R^*_{M_i^*} = A^*_{P_i^*}$. In the above argument we have shown that $R^*_{M_1^*}$, $R^*_{M_2^*}, \ldots,$ are all distinct. Therefore $P_1^*, P_2^*, \ldots,$ are all distinct.

Another proof of part (1) of the above lemma, using valuations, will be given in Chapter II, see Corollary 2.28.

LEMMA 1.29. Let (R, M) be a normal local domain with quotient field K and let (S, N) be a local ring in a finite algebraic extension K' of K lying above R. Then $S \cap K = R$ and $N \cap K = M$.

PROOF. Let K^* be a finite normal extension of K containing K'. Let R' and R^* be the integral closures of R in K' and K^* respectively. Let $M' = N \cap R'$ and let M_1^* be a prime ideal in R^* lying above M'. Then M_1^* is above M. Let $M_2^*, \ldots, M_t^*$ be the other maximal ideals in R^*. Let $S_i^* = R^*_{M_i^*}$, $N_i^* = M_i^* S_i^*$ and let G be the galois group of K^*/K. Now $M' = M_1^* \cap R'$. Therefore $S_1^* \supset S$ and hence $S_1^* \cap K \supset S \cap K$. Hence to show that $S \cap K = R$ it is enough to show that $S_1^* \cap K = R$. Let $u \in S_1^* \cap K$. Given S_i^* there exists $g \in G$ such that $gS_1^* = S_i^*$ and hence $gu \in S_i^*$. But $u \in K$ implies $gu = u$. Hence $u \in S_i^*$ for $i, \ldots, t$, i.e.,

$$u \in \left(\bigcap_{i=1}^{t} S_i^* \right) = R^* ,$$

by 1.10. Therefore $u \in R^* \cap K = R$ and hence $S_1^* \cap K = R$. Therefore $S \cap K = R$ and hence $N \cap K = N \cap R = M$.

DEFINITION 1.30. Let K be a field, K^* a finite algebraic extension of K and S a normal local domain with quotient field K^*. It follows from 1.29 that if there exists a normal local domain R in K such that S lies above R, then R is unique. Hence if R exists, we are justified in calling R "the local ring in K lying below S."

LEMMA 1.31. Let A be a normal domain with quotient field K, P a prime ideal in A, K^* a finite algebraic extension of K, K' a field between K and K^*, A' and A^* the integral closures of A in K' and K^*, respectively. Then A^* is the integral closure of A' in K^*. Let $P_1', \ldots, P_s'$ be the prime ideals in A' lying above P. Let $P_{i1}^*, \ldots, P_{iu_i}^*$ be the prime ideals in A^* lying above P_i^*. Then the $u_1 + \cdots + u_s$ ideals P_{ij}^* are all distinct and they are exactly the prime ideals in A^* lying above P. Let $S_i = A'_{P_i'}$ and $T_{ij} = A^*_{P^*_{ij}}$. Let S_i^* be the integral closure of S_i in K^* and let $N_{i1}^*, \ldots, N_{iv_i}^*$ be the maximal ideals in S_i^*. Then $v_i = u_i$ and after a suitable relabelling we have $T_{ij} = (S_i^*)_{N^*_{ij}}$ for $j = 1, \ldots, u_i$.

PROOF. Exercise in view of what we have proved up to now.

In particular, if $(A, P) = (R, M)$ is a local domain, then we have

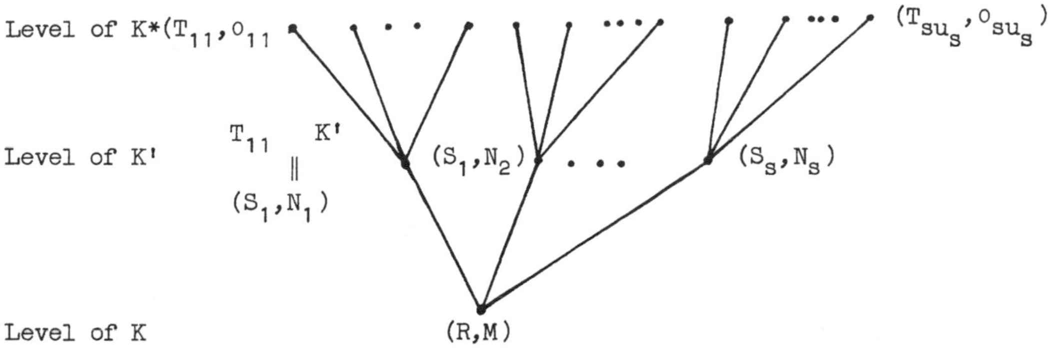

DEFINITION 1.32. Let A be a domain with quotient field K and let P be a prime ideal in A. Let (S, N) be a local domain whose quotient field K^* is an extension of K. If $S \supset A$ and $N \cap A = P$, then we shall say that (S, N) [or S] has center P in A.

LEMMA 1.32A. Now let (R, M) and (S, N) be normal local domains with a quotient field K such that S has center M in R and let (R', M') be a local ring in a finite algebraic extension K' of K lying above R. Then there exists at least one local ring (S', N') in K' lying above S such that $S' \supset R'$. Furthermore, for any such S' we have $N' \cap R' = M'$, i.e., S' has center M' in R'.

PROOF. Let A' and B' be the integral closures of R and S in K' respectively. Let (S', N') be a local ring in K' lying above S such that $S' \supset R'$. Then $[(N' \cap R') \cap A'] \cap R = N' \cap R = (N' \cap S) \cap R = N \cap R = M$ and hence $N' \cap R' = M'$. To prove the existence of S' let K^* be a normal extension of K containing K' and let (S_1^*, N_1^*) be a local ring in K^* lying above S. Let A^* be the integral closures of R in K^*. Then $S_1^* \supset A^*$ and $N_1^* \neq S_1^* \Longrightarrow 1 \notin N_1^* \Longrightarrow N_1^* \cap A^* \neq A^*$. Let P_1^* be a maximal ideal in A^* containing $N_1^* \cap A^*$ and let $R_1^* = A^*_{P_1^*}$ and $M_1^* = P_1^* R_1^*$. Then $S_1^* \supset R_1^*$ and R_1^* is a local ring in K^* lying above R. Let R^* be a local ring in K^* lying above R'. Then R^* is above R and hence by 1.25 there exists a K-automorphism g of K^* such that

$g(R_1^*) = R^*$. Let $g(S_1^*) = S^*$ and let $S' = S^* \cap K'$. Then S' is a local ring in K' lying above S and $S_1^* \supset R_1^* \Longrightarrow S^* \supset R^* \Longrightarrow S' \supset R'$.

LEMMA 1.32B. Let (R, M) (S, N), K, K' be as in 1.32A. Let (S', N') be a local ring in K' lying above S. Then there exists a unique local ring (R', M') in K' lying above R such that S' has center M' in R'.

PROOF. Let A' be the integral closure of R in K'. Then $S' \supset A'$. Let $P' = N' \cap A'$. Then $P' \cap R = (N' \cap A') \cap R = N' \cap R = (N' \cap S) \cap R = N \cap R = M$ and hence P' is a maximal ideal in A'. Hence $R' = A'_{P'}$ is the required unique local ring.

4. Linear algebra. Let A be a ring. An element a of A is called nilpotent if $a^n = 0$ for some positive integer n. An ideal H in A is called a nil ideal if all its elements are nilpotent; H is called a nilpotent ideal if $H^n = \{0\}$ for some positive integer n; obviously a nilpotent ideal is a nil idea. Observe that the set of all the nilpotents in A is an ideal, in fact it is equal to $\sqrt{\{0\}}$. If $\sqrt{\{0\}} = \{0\}$, i.e., if A has no nonzero nilpotents, then A is said to be semisimple. The intersection of all the maximal ideals of A is called the radical of A and is denoted by Rad (A). H is said to be an Artin ring if the descending chain condition for ideals holds in A. A is said to be a primary ring if A contains only one prime ideal.

LEMMA 1.33. Let $S(A)$ denote the set of all the elements a in ring A for which $(1 + xa)$ is a unit for all x in a. Then $S(A) = \text{Rad}(A)$.

PROOF. $a \notin S(A) \Longrightarrow$ there exists x in A such that $(1 + ax)A \neq A \Longrightarrow$ there exists a maximal ideal M in A such that $1 + ax \in M \Longrightarrow ax \notin M \Longrightarrow ax \notin \text{Rad}(A) \Longrightarrow a \notin \text{Rad}(A)$. Therefore $\text{Rad}(A) \subset S(A)$.

$a \notin \text{Rad}(A) \Longrightarrow$ there exists a maximal ideal M in A such that $a \notin M \Longrightarrow M + aA = A \Longrightarrow$ there exist $m \in M$ and $x \in A$ such that $m + ax = 1$, i.e., $1 - ax = m \in M \Longrightarrow 1 - ax$ is a nonunit $\Longrightarrow a \notin S(A)$. Therefore $S(A) \subset \text{Rad}(A)$.

LEMMA 1.34. Let A be an Artin ring. Then $\text{Rad}(A) = \sqrt{\{0\}}$.

PROOF. $a \in \sqrt{\{0\}} \Longrightarrow a^n = 0$ for some $n \Longrightarrow$ for all x in A, $(1 - ax)(1 + ax + a^2x^2 + \dots + a^nx^n) = 1 - a^{n+1}x^{n+1} = 1$ and hence by 1.33, $a \in \text{Rad}(A)$. Hence $\sqrt{\{0\}} \subset \text{Rad}(A)$.

Now let $a \in \text{Rad}(A)$. Since A is an Artin ring, there exists n such that $a^nA = a^{n+1}A$, i.e., there exists x in A such that $a^n = a^{n+1}x$; also by 1.33 there exists y in A such that $(1 - ax)(1 + y) = 1$, i.e., $ax(1 + y) - y = 0$, i.e.,

$0 = a^n(ax)(1 + y) - a^n y = a^n(1 + y) - a^n y = a^n$; therefore $a \in \sqrt{\{0\}}$. Hence $\mathrm{Rad}(A) \subset \sqrt{\{0\}}$.

LEMMA 1.35. In an Artin ring A every nilideal H is nilpotent.

PROOF. Since A is Artin, there exists n such that $H^n = H^{n+1} = H^{n+2} = \ldots = I$, say. Suppose, if possible, that $I \neq 0$ and let W be the set of all ideals J in A such that $JI \neq 0$. Since A is Artin, W contains a minimal element K. Then there exists k in K such that $kI \neq 0$ so that $kA \in W$. Since $kA \in W$ and $kA \subset K$ we must have $kA = K$. Now $KI^2 = KI \neq 0$ and hence $KI \in W$. Since $KI \subset K$ we must have $kI = KI = K = kA$. Therefore there exists i in I such that $k = ki$ and hence $k = ki = ki^2 = \ldots = ki^n = \ldots$. Since i is nilpotent, we must have $k = 0$, i.e., $KI = 0$, which is a contradiction. Therefore $I = 0$.

LEMMA 1.36. Let A be a primary Artin ring. Then every zero divisor in A is nilpotent, i.e., belongs to $\sqrt{\{0\}} = \mathrm{Rad}\,(A)$, and $\mathrm{Rad}\,(A)$ is the unique prime ideal in A. If $\mathrm{Rad}\,(A) = \{0\}$ then A is a field and conversely.

PROOF. Straightforward.

PROPOSITION 1.37. Let A be an Artin ring. Then

(1) A contains only a finite number of prime ideals $P_1, P_2, \ldots, P_m$.

(2) Each P_i is maximal.

(3) A has a unique direct sum decomposition $A = B_1 \oplus \cdots \oplus B_n$ where the B_i are ideals each being a primary ring and we have $n = m$.

(4) Let e_i be the identity in B_i. Then after a suitable rearrangement we get

$$e_i = \begin{cases} 1 \bmod P_i \\ 0 \bmod P_j \quad \text{if } i \neq j \end{cases}.$$

(5) Let N and N_i be the radicals of A and B_i respectively. Then $N_i = Ne_i = P_i e_i$; $N = N_1 \oplus \cdots \oplus N_m$.

(6) Every chain of ideals in A is finite and hence in particular A is noetherian.

(7) If A is semisimple, then each B_i is a field and $B_1, \ldots, B_m$ are exactly the nonzero minimal ideals of A.

(8) If A is an algebra over a field K and if $\{a_{ij}\}$,

$j = 1, \ldots, n_i$, is a $K(= Ke_i)$-basis of the algebra B_i, then $\{a_{ij}\}$ $(i = 1, \ldots, m;\ j = 1, \ldots, n_i)$ is a K-basis of A and hence $[A : K] = \Sigma[B_i : K(= Ke_i)]$. Finally, if A is semisimple, then B_i is an overfield of $K(= Ke_i)$ [Theorem of Dedekind].

PROOF. Let N be the radical of A. By 1.34 and 1.35, N is the set of all nilpotents in A and there exists s such that $N^s = 0$. Since A is Artin, A has only a finite number of maximal ideals $P_1, \ldots, P_m$ and by 1.2 and 1.3 we have

$$N = \bigcap_{i=1}^{m} P_i = \prod_{i=1}^{s} P_i; \qquad N^s = \prod_{i=1}^{s} P_i^s = \bigcap_{i=1}^{s} P_i^s \ .$$

For $i \neq j$, $P_j \not\subset P_i$ and hence there exists $x_{ij} \in P_j, \notin P_i$. Let

$$x_i = \prod_{j \neq i} x_{ij} \ .$$

Then $x_i \notin P_i$ and $x_i \in P_j$ if $j \neq i$. Since A/P_i is a field, there exists c_i in A such that $c_i x_i \equiv 1(P_i)$. Let $e_i = 1 - [1 - (c_i x_i)^s]^s$. Then

$$e_i \equiv \begin{cases} 1 \bmod P_i^s \\ 0 \bmod \prod_{j \neq i} P_j^s \ . \end{cases}$$

For $i \neq j$,

$$e_i e_j \in \prod_{k=1}^{m} P_k^s = N^s = 0 \ ,$$

i.e., $e_i e_j = 0$. Also

$$e_i(1 - e_i) \in \prod_{k=1}^{m} P_k^s = N^s = 0 \ ,$$

i.e., $e_i(1 - e_i) = 0$, i.e., $e_i^2 = 1$. Finally

$$(e_1 + \ldots + e_m - 1) \in \bigcap_{k=1}^{m} P_k^s = N^s = 0 \ ,$$

i.e., $e_1 + \dots + e_m = 1$.

Let $B_i = e_iA$. Then we have the direct sum decomposition $A = B_1 \oplus \dots \oplus B_m$ which is also a direct product decomposition. Since for $j \neq 1$, $e_1 \in P_j$; e_1P_j contains $e_1^2 = e_1$ which is the identity in B_1 and hence

$$\bigcap_{k=1}^{m} P_k e_1 = P_1 e_1 \quad .$$

Since for $j \neq 1$, $e_j \in P_1$, we have $P_1 \supset B_2 \oplus \dots \oplus B_m$ = kernel f_1 where f_1 is the homomorphism of A onto B_1 given by: $f_1(a) = ae_1$. Therefore $N_1 = P_1e_1$ is a maximal ideal in B_1. Since $N_1 = Ne_1$ is the set of all nilpotents of B_1 and since A is Artin implies B_1 is Artin, by 1.34 we conclude that N_1 is the only maximal ideal of B_1. Now if P is a prime ideal in B_1 then N_1 = (the set of all nilpotents in B_1) $\subset P$ which implies $P = N_1$. Therefore N_1 is the only prime ideal in B_1. Similarly $N_i = P_ie_i$ is the only prime ideal in B_i and N_i is the set of all nilpotents in B_i. Therefore B_i is primary. Again P is a prime ideal in $A \Longrightarrow N \subset P \Longrightarrow P_1P_2 \dots P_m \subset P \Longrightarrow P_i \subset P$ for some i, i.e., $P = P_i$. That $N = N_1 \oplus \dots \oplus N_m$ is now obvious, in fact for any ideal H we have $H = H_1 \oplus \dots \oplus H_m$ where $H_i = He_i$.

Now let $A = D_1 \oplus \dots \oplus D_n$ be a direct sum decomposition of A into nonzero ideals D_i. Then $1 = d_1 + \dots + d_n$ with $d_i \in D_i$, $d_i^2 = d_i$, $d_id_j = 0$ for $i \neq j$ and $D_i = d_iA$. Also $M_i = Nd_i$ = radical of D_i. Assume $D_1, \dots, D_n$ are primary. Then M_i is the unique maximal ideal in D_i. Now

$$\begin{aligned} d_1 &= d_1e_1 + \dots + d_1e_m \\ &= d_1^2 = d_1^2e_1 + \dots + d_1^2e_m \quad . \end{aligned}$$

Therefore d_1e_i is either zero or is an idempotent in B_i. Since in B_i every nonnilpotent is a unit, $d_1e_i = 0$ or e_i. Therefore either $B_i \subset D_1$ or $B_i \cap D_1 = \{0\}$. Hence after a suitable reordering we have: $D_1 = B_1 \oplus \dots \oplus B_s$. Similarly after a suitable reordering: $B_1 = D_1 \oplus \dots \oplus D_t$. Therefore $B_1 = D_1$. Hence $n = m$ and after a suitable reordering we have: $B_i = D_i$ for $i = 1, \dots, m$. Thus we have proved statements (1) to (5). To prove (6), it is enough to prove it for each B_i, i.e., to prove that in a primary Artin ring B every chain of ideals is finite. Let M be the radical and hence the only prime ideal of B. If $M = 0$ there is nothing to show. So assume $M \neq \{0\}$, and let n be such that $M^{n-1} \neq 0$ and $M^n = 0$. Consider the B-modules $B/M, M/M^2, \dots, M^{n-1}/M^n$. Since each of them is annulled by M, they may be considered to be modules over the field B/M. They are Artin rings and vector spaces and hence they are finite dimensional vector spaces. Therefore there are only a finite number of B-modules, i.e., B-ideals between M^i and M^{i-1} for $i = 1, \ldots, n$. Therefore by the Jordan-Hoelder theorem every chain of ideals in B is finite. This proves (6). Now let us prove (7). $N_i = 0$ implies B_i is a field which implies that B_i is a (nonzero) minimal ideal. Now let P be any other (nonzero) minimal ideal in A. Then $Pe_i = 0$ or B_i and hence $P = B_i$ for some i. Finally, now (8) is obvious.

5. Discriminant of an algebra. Let A be a ring, and K a subfield of A such that A is a finite-dimensional vector space over K, say $[A : K] = n$. Let $(w) = (w_1, \ldots, w_n)$ be a basis of A/K. Given a in A let T_a be the linear transformation of the vector space A into itself given by $T_a(b) = ab$ for $b \in A$. Let

$$aw_i = \sum_{j=1}^{n} k_{ij}w_j$$

with $k_{ij} \in K$ so that $((k_{ij}))$ is the matrix of T_a with respect to the basis $(w_1, \ldots, w_n)$. Define the characteristic polynomial of a with respect to A/K by:

$$\begin{aligned} Ch_a(X) = Ch_{A/K,\, a}(X) &= \det\,(XE - ((k_{ij}))), \quad [E = \text{unit matrix}] \\ &= X^n + c_1X^{n-1} + \ldots + c_n \quad . \end{aligned}$$

Also define:

$$S(a) = S_{A/K}(a) = (\text{trace of } a \text{ relative to } A/K) = -\, c_1 = \sum_{i=1}^{n} k_{ii} \quad ,$$

$$N(a) = N_{A/K}(a) = (\text{norm of } a \text{ relative to } A/K) = (-\,1)^n c_n = \det\,((k_{ij})) \; .$$

Let $(w^*) = (w_1^*, \ldots, w_n^*)$ be any other basis of A/K and let L be the matrix: $(w^*) = L(w)$. Then the matrix of T_a relative to the basis (w^*) is: $((k_{ij}^*)) = L^{-1}((k_{ij}))L$ and hence

$$\begin{aligned} \det\,(XE - ((k_{ij}))) &= \det\,(L^{-1}(XE - ((k_{ij}))L) \\ &= \det\,(XE - ((k_{ij}^*))) \quad . \end{aligned}$$

Thus $Ch_a(X)$ and hence $S(a)$ and $N(a)$ do not depend on the particular choice of the basis of A/K.

Let $a, b \in A$. Let

$$T_a : w_i \longrightarrow aw_i = \sum k_{ij}w_j$$

$$T_b : w_i \longrightarrow bw_i = \sum \ell_{ij}w_j \quad .$$

Then

$$T_{ab} = T_aT_b : (w) \longrightarrow ((k_{ij}))(\ell_{ij}))(w) \quad .$$

Therefore

$$N(ab) = \det(((k_{ij}))((\ell_{ij}))) = \det\ ((k_{ij}))\det\ ((\ell_{ij})) \qquad (1)$$
$$= N(a)N(b)$$

$$S(a + b) = \sum (k_{ii} + \ell_{ii}) = \sum k_{ii} + \sum \ell_{ij} \qquad (2)$$
$$= S(a) + S(b)$$

Let $k \in K$. Then

$$kaw_i = \sum (kk_{ij})w_j \quad .$$

Therefore

$$S(ka) = kS(a) \quad \text{and} \quad N(ka) = k^nN(a) \qquad (3)$$

$T_1 : w_i \longrightarrow w_i$. Hence $S(1) = n$ and $N(1) = 1$. Therefore by (3),

$$S(k) = nk \quad \text{and} \quad N(k) = k^n \quad . \qquad (4)$$

Let $(a_1, \ldots, a_n)$ be an arbitrary set of elements of A. We define:

$D(a_1, \ldots, a_n) = D_{A/K}(a_1, \ldots, a_n) =$ (discriminant of $(a_1, \ldots, a_n)$ relative to $A/K) = \det ((S(a_ia_j)))$.

Let $(b_1, \ldots, b_n)$ be another set of elements of A with $b_i = \Sigma k_{ij}a_j$, $k_{ij} \in K$. Then

$$b_vb_u = \left(\sum_i k_{vi}a_i\right)\left(\sum_j k_{uj}a_j\right) = \sum_{i,j} k_{vi}a_ia_jk^*_{ju} \quad ,$$

where $k^*_{ju} = k_{uj}$. Therefore

$$((S(b_{vu})))^{\cdot} = \left(\left(S\left(\sum_{i,j} k_{vi}a_ia_jk^*_{ju}\right)\right)\right)$$

$$= \left(\left(\sum_{i,j} k_{vi}S(a_ia_j)k^*_{ju}\right)\right)$$

$$= ((k_{ij}))((S(a_ia_j)))((k^*_{ij})) \quad .$$

Therefore

$$D(b_1, \ldots, b_n) = [\det ((k_{ij}))]^2 \, D(a_1, \ldots, a_n) \quad . \qquad (5)$$

Let now $(w^*) = L(w)$ be any other basis of A/K. Then $D(w^*) = D(w^*_1, \ldots, w^*_n) = [\det (L)]^2 D(w)$. Since $\det (L) \neq 0$, $D(w^*) = 0 \Longleftrightarrow D(w) = 0$. Hence, though D depends on the basis, it makes sense to say that

"the discriminant of A/K is zero (or nonzero)" (6)

Now let $A = A_1 \oplus \cdots \oplus A_m$ be a direct product decomposition of A, let e_i be the unit of A_i, let $K_i = Ke_i$, and (w_{ij}) $(j = 1, \ldots, n_i)$ be a basis of A_i/K_i. Then $\{w_{ij}\}$ $(i = 1, \ldots, m;\ j = 1, \ldots, n_i)$ is a basis of A/K. For $a \in A$, $a = ae_1 + \cdots + ae_m$.

$$(ae_i)w_{ij} = \sum_{u=1}^{n_i} k^{(i)}_{ju} e_i w_{iu} \qquad (\text{with } k^{(i)}_{ju} \in K)$$

$$= \sum_{u=1}^{n_i} k^{(i)}_{ju} w_{iu} \quad .$$

Hence

$$aw_{ij} = \left(\sum_i ae_i\right) w_{ij} = ae_iw_{ij} = \sum_{u=1}^{n_i} k^{(i)}_{ju} w_{iu} \quad .$$

Therefore

$$S_{A/K}(a) = S_{A_1/K_1}(ae_1) + \cdots + S_{A_m/K_m}(ae_m) \quad ,$$

$$N_{A/K}(a) = \prod_{i=1}^{m} N_{A_i/K_i}(ae_i) \quad ,$$

and $i \neq u \Longrightarrow w_{ij}w_{uv} = 0 \Longrightarrow S(w_{ij}w_{uv}) = 0$. Therefore

$$D_{A/K}(w_{11}, w_{12}, \ldots, w_{1n_1}, w_{21}, \ldots, w_{mn_m})$$

$$= \det ((S_{A/K}(w_{ij}w_{uv})))$$

$$= \prod_{i=1}^{m} \det ((S_{A/K}((w_{ij}w_{iv}))) \qquad [j, v = 1, \ldots, n_i] \quad ,$$

and

$$D_{A_i/K_i}(w_{i1}, \ldots, w_{in_i}) = \det ((S_{A_i/K_i}(w_{ij}w_{iv})))$$

$$= \det ((S_{A/K}((w_{ij}w_{iv})))e_i \quad .$$

Therefore

The discriminant of A/K is zero $\Longleftrightarrow$ the discriminant of A_i/K_i is zero for $i = 1, \ldots, m$. (7)

Now suppose that the radical N of A is nonzero, i.e., A has nonzero nilpotents, and let s be such that $N^{s-1} \neq 0$ and $N^s = 0$ $(s > 1, N^0 = A)$. Then $A = N^0 > N^1 > \ldots > N^s = \{0\}$. N^i is a vector space over K and hence we can find elements $p_{i1}, \ldots, p_{iq_i}$ in N^i whose residue classes modulo N^{i+1} form a K-basis of the vector space N^i/N^{i+1} and then $\{p_{ij}\}$ $(i = 0, 1, \ldots, s - 1; j = 1, \ldots, q_i)$ is a basis of A/K. Now

$$a \in N \Longrightarrow ap_{ij} = \sum_{v>i} k_{ijvu}p_{vu} \; (k_{ijvu} \in K) \Longrightarrow S(a) = 0 \quad . \tag{8}$$

$N \neq 0 \Longrightarrow$ there exists a basis $w_1, \ldots, w_n$ of A/K with $w_n \in N \Longrightarrow$ discriminant of A/K is zero . (9)

Recall from field theory that in the special case when A is a field we have: For a given $a \in A$ let $a = a^{(1)}, \ldots, a^{(n)}$ be the conjugates of a (counted with proper multiplicities). Then

$S(a) = a^{(1)} + \dots + a^{(n)}$, $N(a) = a^{(1)}a^{(2)} \dots a^{(n)}$, $D(a_1, \dots, a_n) = \det((a_i^{(j)}))^2$ and discriminant of A/K is zero if and only if A/K is inseparable. (10)

Now assume that $A_1, \dots, A_m$ are primary rings, i.e., $A = A_1 \oplus \dots \oplus A_m$ is the unique decomposition of A referred to in 1.37. Then N_i is the unique prime ideal in A_i and A_i/N_i is a field. We define

$$[A_i : K]_s = (\text{the separable degree of } A_i \text{ over } K) = [(A_i/N_i) : K_i]_s \tag{11}$$

If $(A_i/N_i)/K_i$ is separable for $i = 1, \dots, m$ then we shall say that

$$A \text{ is separable over } K \quad . \tag{12}$$

Now (7), (9) and (10) yield

PROPOSITION 1.38. A/K has a nonzero discriminant $\Longleftrightarrow$ A/K is semisimple and separable

$$\Longleftrightarrow [A : K] = \sum_{i=1}^{m} [Ai : K]_s \quad .$$

Observe that any element $a \in A$ satisfies a polynomial equation $f(a) = 0$ where $f(X)$ is a nonzero monic polynomial in $K[X]$ of degree at most n. If a satisfies no equation of degree less than n, i.e., if $1, a, \dots, a^{n-1}$ is a basis of A/K, then we say that a is a <u>primitive element</u> of A/K. Observe that if a is a primitive element of A/k and if $f(X) = X^n + k_1X^{n-1} + \dots + k_n$ is the unique monic polynomial of degree n in $K[X]$ such that $f(a) = 0$, then we have

$$D_{A/K}(a) = D_{A/K}(1, a, \dots, a^{n-1}) = D_n^*(k_1, \dots, k_n) \quad ,$$

where $D_n^*(X_1, \dots, X_n)$ is an element of $Z[X_1, \dots, X_n]$ completely determined by n where Z is the ring of integers (in fact D^* is the discriminant of f in the elementary sense).

Lemma 1.39. In the above notation, suppose that A/K is semisimple separable and assume that K is infinite. Then A/K has primitive elements.

In view of 1.38, the proof of this lemma follows from the next more general lemma.

LEMMA 1.40. Let K be an infinite field and let A be an overring of K. Let $A_1, \ldots, A_m$ be nonzero ideals in A such that $A = A_1 \oplus \cdots \oplus A_m$. Let e_i be the identity in A_i and let $K_i = Ke_i$. Assume that A_i is a primary ring in which the unique prime ideal N_i consists entirely of nilpotents (i.e., N_i is the set of all the nilpotents of A_i). Let $q_1, \ldots, q_m$ be positive integers such that $q_i \leqq [(A_i/K_i) : K_i]_s$. Then (1) there exists a in A such that a does not satisfy any equation in $K[X]$ of degree $< q_1 + \cdots + q_m$; (2) assume furthermore that for some j either $N_j \neq (0)$ or that $[(A_j/N_j) : K_j]_i > 1$. Then there exists a in A such that a does not satisfy any equation in $K[X]$ of degree $\leqq q_1 + \cdots + q_m$. [If $[A : K] = n < \infty$ then we can state this as: (1) Lemma 1.39; and (2) if A/K is not semisimple separable, then there exists a in A such that a does not satisfy any equation in $K[X]$ of degree $\leqq \Sigma_{i=1}^{m} [A_i : K]_s$.]

PROOF. For c_i in A_i we shall denote by $\bar{c}_i$ the N_i-residue class of c_i.

Observe that since any zero divisor in N_j is necessarily a nilpotent we have the following: For a_j in A_j let H be the ideal of polynomials $f(X)$ in $K_j(X)$ such that $f(a_j) = 0$, then $H = u(X)^r K_j[X]$ where $u(X)^r$ is the <u>monic minimal polynomial</u> of a_j/K_j and it is a power r of $u(X)$ which is irreducible and is the minimal monic polynomial of $\bar{a}_j$ over K_j.

We are going to choose an element a_j in A_j for $j = 1, \ldots, m$. Fix b_j in A_j such that the monic minimal polynomial $h_j(X)$ of $\bar{b}_j$ over K_j is of degree $s_j \geqq q_j$. If $N_j = 0$ and $[A_j : K_j]_i = 1$, then we set $a_j = b_j$ and $g_j^*(X) =$ the minimal monic polynomial of a_j over K_j. Then $g_j^*(X) = h_j(X)$ and $\deg g_j^*(X) = t_j = s_j \geqq q_j$. Now assume that either $N_j \neq 0$ or $[(A_j/N_j) : K_j]_i > 1$. We want to show then that there exists a_j in A_j whose minimal monic polynomial $g_j^*(X)$ over K_j is of degree $t_j > q_j$. If $h_j(b_j) \neq 0$ we may set $a_j = b_j$, so assume that $h_j(b_j) = 0$.

CASE 1. $[(A_j/N_j) : K_j]_i > 1$. Then by the theorem of primitive elements [W1, p. 126] there exists a_j in A_j such that the minimal monic polynomial of $\bar{a}_j$ over K_j is of degree $\geqq s_j + p > q_j$ (where p is the characteristic of K_j).

CASE 2. $[(A_j/N_j) : K_j]_i = 1$. Then there exists $0 \neq c_j \in N_j$. Let $a_j = b_j + c_j$. Then $h_j'(b_j)^r \neq 0$ for all r (here $h_j'(X)$ is the derivative of $h_j(X)$), i.e., $h_j'(b_j) \notin N_j$. Now

$$h_j(a_j) = h_j(b_j + c_j) = c_j[h_j'(b_j) + c_j f(b_j)]$$

with $f(X) \in K_j[X]$, $h_j'(b_j) \notin N_j$ and $c_j f(b_j) \in N_j \Longrightarrow h_j'(b_j) + c_j f(b_j) \notin N_j \Longrightarrow h_j'(b_j) + c_j f(b_j)$ is a nonzero divisor $\Longrightarrow h_j(a_j) \neq 0$. Since

$h_j(a_j)^r = 0$ for some r and since $h_j(X)$ is prime, for the minimal monic polynomial $g_j^*(X)$ of a_j/K_j we must have: $g_j^*(X) = h_j(X)^r$, with $r > 1$, i.e., $t_j = \deg g_j^*(X) > q_j$. Thus the existence of a_j has been established.

Now let $g_j(X) = X^{t_j} + k_{j1}X^{t_j-1} + \dots + k_{jt_j}$ be the monic polynomial in $K[X]$ such that

$$e_jX^{t_j} + e_jk_{j1}X^{t_j-1} + \dots + e_jk_{jt_j} = g_j^*(X) .$$

Multiplying the a_j by suitable elements of the infinite fields K_j we may assume that $g_1(X), \dots, g_m(X)$ are pairwise coprime. Let $a = a_1 + \dots + a_m$. Then

$$f(X) = k_oX^q + \dots + k_q \in K[X] \quad \text{with } f(a) = 0 \Longrightarrow$$

$$f^{(j)}(a_j) = 0, \quad \text{where } f^{(j)}(X) = e_jk_oX^q + \dots + e_jk_q \Longrightarrow$$

$$g_j^*(X) \text{ divides } f^{(j)}(X) \text{ in } K_j[X] \text{ for } j = 1, \dots, m \Longrightarrow$$

$$g_j(X) \text{ divides } f(X) \text{ in } K[X] \text{ for } j = 1, \dots, m .$$

Therefore for the degree d of the minimal monic polynomial of a over K we have:

$$d \geq t_1 + \dots + t_m \begin{cases} = q_1 + \dots + q_m \\ > q_1 + \dots + q_m, \text{ if for some } j \text{ we have} \\ \quad N_j \neq 0 \text{ or } [(A_j/N_j) : K_j] > 1 . \end{cases}$$

6. <u>The discriminant ideal</u>. Let A be a normal domain with quotient field K. Let K^* be a finite algebraic extension of K and let A^* be a subdomain of K^* with K^* as quotient field such that A^* contains A and is integral over A, let $[K^* : K] = n$. Let $(w_1', \dots, w_n')$ be a basis of K^*/K. Since K^* is the quotient field of A^*, we can write $w_i' = u_i/v_i$ with $u_i, v_i \in A^*$ and $v_i \neq 0$. Let

$$w_i = w_i' \prod_{j=1}^{n} v_j .$$

Then $(w_1, \dots, w_n)$ is a basis of K^*/K and $w_i \in A^*$ for $i = 1, \dots, n$. Then $w_iw_j \in A^*$ and hence $S_{K^*/K}(w_iw_j) \in A$ since A^*/A is integral. Therefore $D_{K^*/K}(w_1, \dots, w_n) \in A$. We define

$D(A^*/A)$ = (the discriminant ideal of A^* over A)
= the ideal in A generated by the discriminants of all the bases of K^*/K which are in A^*.

Now let A' be the integral closure of A in K^*. We define

$$D(K^*/A) = (\text{the discriminant of } K^* \text{ over } A) = D(A'/A) .$$

Now let P be a prime ideal in A, then by 1.26 there are only a finite number of prime ideals $P_1^*, \ldots, P_m^*$ in A^* lying above P. Let $S_j = A^*_{P_j^*}$ and $N_j = P_j^* S_j$. Then $A^*/P_j^* = S_j/N_j \supset A/P$. If for a given element of A^* we reduce modulo P the equation of integral dependence over A, we obtain an equation of integral dependence for the corresponding residue class modulo P_j over A/P. Therefore $A^*/P_j^* = S_j/N_j$ is an algebraic extension of A/P.

Now assume that A is a local ring with P as its maximal ideal. Then $(A^*; P_1^*, \ldots, P_m^*)$ is a semilocal ring. Let P^* be any prime ideal in A^* such that $P^* \supset PA^*$. Since $P_1^*, \ldots, P_m^*$ are all the maximal ideals of A^*, we must have $P^* \subset P_j^*$ for some j. Now $P \subset P^*$ and hence $P \subset P^* \cap A \subset P_j^* \cap A = P$, so that $P^* \cap A = P$ and hence $P^* = P_j^*$ by 1.19. Hence $P_1^*, \ldots, P_m^*$ are the only prime ideals in A^* which contain PA^*. Therefore by 1.11 we can uniquely write $PA^* = Q_1^* \cap \ldots \cap Q_m^*$, where Q_j^* is primary for P_j^*. By 1.9, $PS_j = Q_j^* S_j$ and $Q_j^* = Q_j^* S_j \cap A^* = PS_j \cap A^*$ and $PS_j = N_j$ if and only if $Q_j^* = P_j^*$. Since PS is primary for N_j, S_j/PS is a primary ring whose unique prime ideal N_j/PS coincides with its ideal of nilpotents and it is semisimple if and only if $PS = N_j$. Thus the following three conditions are equivalent:

(1) $PS_j = N_j$ and $(S_j/N_j)/(A/P)$ is separable;

(2) (S_j/PS_j) is semisimple separable over (A/P);

(3) $Q_j^* = P_j^*$ and $(A^*/P_j^*)/(A/P)$ is separable.

If one, hence all three, conditions are satisfied, we shall say that S_j (respectively N_j or P_j^*) is <u>unramified over</u> A (respectively over P). If S_j is not unramified over A then we shall say that S_j (respectively N_j or P_j^*) is <u>ramified over</u> A (respectively over P). If $S_1, S_2, \ldots, S_m$ <u>are all unramified over</u> A, we shall say that A^* is <u>unramified over</u> A; otherwise we shall say that A^* is ramified over A.

Now let us get back to the general case when A is not necessarily local and assume that A^* is the integral closure of A in K^*. Let $R = A_P$ and $M = PR$. Let R^* be the integral closure of R in K^*. Let $M_j^* = R^* \cap N_j$. Then by 1.28, $M_1^*, \ldots, M_m^*$ are exactly the maximal ideals in R^*, $S_j = R^*_{M^*_j}$ and $N_j = M^*_jS_j$. We shall say that P^*_j (or N_j or S_j) is ramified or unramified over P (or A) according as S_j is ramified or unramified over R. We shall say that P is a <u>branch ideal</u> or a <u>nonbranch ideal</u> of A for the extension A^*/A (or for the extension K^*/K, or for $K \longrightarrow K^*$) according as R^* is ramified or unramified over R.

LEMMA 1.41. Let (R, M) be a normal local domain with quotient field K, let K' be a finite algebraic extension of K and let K^* be a finite algebraic extension of K'. Let (R', M') be a local ring in K' lying above R and let (R^*, M^*) be a local ring in K^* lying above R'. (1) If $MR' = M'$ then $MR^* = M^* \Longleftrightarrow M'R^* = M^*$; (2) If R' is unramified over R then R^* is unramified over $R \Longrightarrow R^*$ is unramified over R'.

PROOF. Obvious.

Lemma 1.41A. Let the notation be as in 1.41. Assume that R contains a subfield k and a finite number of elements $a_1, \ldots, a_t$ such that $R = A_P$ where $A = k[a_1, \ldots, a_t]$ and $P = A \cap M$. Then (2a) If R^* is unramified over R, then R' is unramified over R.

The proof of this lemma is rather involved, and hence we refer to the considerations of Section 2 of [A2] from which it follows directly.

Now we are ready to prove the discriminant theorem in its local form.

THEOREM 1.42. Let (R, M) be a normal local domain with quotient field K, let K^* be a finite algebraic extension of K, let R^* be a domain with quotient field K^* such that R^* contains and is integral over R. Let $M^*_1, \ldots, M^*_t$ be the maximal ideals in R^*. Then

$$\sum_{j=1}^{t} [(R^*/M^*_j) : (R/M)]_s \leq [K^* : K] \quad ,$$

and the equality holds if and only if $D(R^*/R) = R$ [i.e., if and only if there exists a basis $(w_1, \ldots, w_n)$ of K^*/K in R^* such that $D_{K^*/K}(w_1, \ldots, w_n)$ is a unit in R (since M is the ideal of all the nonunits of R)].

PROOF. We shall give a proof under the assumption that R/M is infinite. For an extension of the proof for finite R/M see [K1]. Let $[K^* : K] = n$ and $[(R^*/M^*_j) : (R/M)]_s = g_j$. Let $N^* = MR^* = N^*_1 \cap \ldots \cap N^*_t$ where N^*_j is primary for M^*_j. Let $A = R^*/N^*$ and $A_j = R^*/N^*_j$. Then A and A_j can be supposed to contain the field $k = R/M$. Let $Q_j = N^*_j/N^*$ and $P_j = M^*_j/N^*$. Then in A the zero ideal has the decomposition

$\{0\} = Q_1 \cap \dots \cap Q_t$ into pairwise coprime primary ideals Q_j. Hence $A = A_1 \oplus \dots \oplus A_m$. Since Q_j is primary for P_j, it follows that the unique prime ideal P_j in A_j coincides with the ideal of nilpotents of A_j. It follows by 1.16 that every element of A satisfies an equation of degree at most n over k. Hence by 1.40,

$$\sum_{j=1}^{t} [(R^*/M_j^*) : (R/M)]_s = \sum_{j=1}^{t} [A_j : k]_s \leq n \quad .$$

Let us agree to denote for a in R^* the residue class modulo N^* by $\bar{a}$.

Now assume that $g_1 + \dots + g_t = n$. Then by 1.40 and 1.16, A_j has no nonzero nilpotents [hence $N_j^* = M_j^*$], i.e., A_j is a field and it is separable over k, i.e., $[A_j : k] = [A_j : k] = g_j$. Hence A_j is a finite-dimensional algebra over k for $j = 1, \dots, t$. Hence A is a finite-dimensional algebra over k and hence by 1.38 A/k has a nonzero discriminant. Therefore by 1.39 we can find a primitive element u of A/k. Then $0 \neq D(1, \bar{u}, \dots, \bar{u}^{n-1})$ = the residue class modulo M of $D(1, u, \dots, u^{n-1})$ and hence $D(1, u, \dots, u^{n-1}) \notin M$, i.e., it is a unit in R and hence $D(R^*/R) = R$.

Now assume that $D(R^*/R) = R$. Then there exists a basis $(w_1, \dots, w_n)$ of K^*/K in R^* such that $D(w_1, \dots, w_n)$ is a unit in R. Certainly $R^* \supset w_1R + \dots + w_nR$. We now show that $w_1R + \dots + w_nR$ = the integral closure of R in K^*. Given w in K^* we have $w = a_1w_1 + \dots + a_nw_n$ with $a_j \in K$. Also $w \notin w_1R + \dots + w_nR$ implies some $a_j \notin R$, say $a_1 \notin R$; then $D(w, w_2, \dots, w_n) = a_1^2 D(w_1, w_2, \dots, w_n) \notin R \implies w/R$ not integral. Therefore $R^* = w_1R + \dots + w_nR$ = the integral closure of R in K^*. Therefore $(\bar{w}_1, \dots, \bar{w}_n)$ is a k-basis of A and $N^* = w_1M + \dots + w_nM$. Let

$$(w_i w_j) w_p = \sum_{q=1}^{n} a_{ijpq} w_q, \qquad \text{with} \qquad a_{ijpq} \in R \quad .$$

Then

$$(\bar{w}_i \bar{w}_j) \bar{w}_p = \sum_{q=1}^{n} \bar{a}_{ijpq} \bar{w}_q \quad .$$

Therefore

$D(\bar{w}_1, \dots, \bar{w}_n)$ = the residue class of $D(w_1, \dots, w_n)$ modulo M.

Hence $D(\bar{w}_1, \ldots, \bar{w}_n) \neq 0$. Therefore, by 1.38, $g_1 + \cdots + g_t = n$.

The above proof yields the following two results.

PROPOSITION 1.43. Let the notation be as in 1.42. Then $D(R^*/R) = R \Longrightarrow R^*$ is normal (i.e., R^* is the integral closure of R in K^*) and has an R-module basis of $[K^* : K]$ elements (which are therefore linearly independent over K), i.e., R^* is a free R-module.

THEOREM 1.44. Let the notation be as in 1.42. Then $D(R^*/R) = R \Longrightarrow R^*$ is unramified over R.

We remark that the converse of 1.44 is not true in general (see [K1]); however, it holds under certain assumptions, namely we have

THEOREM 1.44A. Let A be a finitely generated domain over a field and P a prime ideal in A. Let $R = A_P$, $M = PR$, K = the quotient field of R, K^* = a finite algebraic extension of K and R^* = the integral closure of R in K^*. Then R^* is unramified over $R \Longrightarrow D(R^*/R) = R$.

For the proof of this theorem, we refer to its original source [A2, Section 2].

We can globalize Theorem 1.42 as follows:

THEOREM 1.45. Let A be a normal domain with quotient field K, let K^* be a finite algebraic extension of K and let P be a prime ideal in A. Let A^* be the integral closure of A in K^* and let $P_1^*, \ldots, P_t^*$ be the prime ideals in A^* lying above P. Then

$$\sum_{j=1}^{t} [(A^*/P_j^*) : (A(P)]_s \leqq [K^* : K]$$

and the equality holds if and only if $D(A^*/A) \not\subset P$.

PROOF. In view of 1.42 and 1.28, it is enough to show that $D(A^*/A) \not\subset P \Longleftrightarrow D(R^*/R) = R$ where $R = A_P$, $M = PR$ and R^* = the integral closure of R in K^*. ($\Longrightarrow$) $D(A^*/A) \not\subset P$ implies that there exists a basis $(w_1, \ldots, w_n)$ of K^*/K in A^* such that $D(w_1, \ldots, w_n) \notin P$. Then $w_j \in R^*$ and $D(w_1, \ldots, w_n)$ is a unit in R and hence $D(R^*/R) = R$. ($\Longleftarrow$) $D(R^*/R) = R$ implies that there exists a basis $(w_1, \ldots, w_n)$ of K^*/K in R^* such that $D(w_1, \ldots, w_n)$ is a unit in R. Fix $a_{ij} \in R$ such that

$$w_j^{m_j} + a_{j1}w_j^{m_j-1} + \cdots + a_{jm_j} = 0 \quad .$$

Let $a_{ij} = b_{ij}/c_{ij}$ with $b_{ij}, c_{ij} \in A$ and $c_{ij} \notin P$. Let

$$c_j = \prod_{i=1}^{m_j} c_{ji} \quad .$$

Then $w_j c_j \in A^*$ and

$$D(w_1 c_1, \ldots, w_n c_n) = \left(\prod_{j=1}^{n} c_j \right)^2 D(w_1, \ldots, w_n)$$

$$= \text{a unit in } R \quad ,$$

and hence

$$D(w_1 c_1, \ldots, w_n c_n) \in A \text{ and } \notin P \quad .$$

7. Galois theory of local rings. Let A be a normal domain with quotient field K, let K^* be a galois (i.e., finite normal separable algebraic) extension of K, let P be a prime ideal in A, let $A^* =$ the integral closure of A in K^*. Let $R = A_P$, $M = PR$, and let $(R_1^*, M_1^*), \ldots, (R_u^*, M_u^*)$ be the local rings in K^* lying above R. Let $P_j^* = M_j^* \cap A^*$. Let G be the galois group of K^*/K. Observe that for $g \in G$ and a given j the following conditions are equivalent:

(1) $g(R_j^*) = R_j$;

(2) $g(M_j^*) = M_j^*$;

(3) $g(P_j^*) = P_j^*$

(proof obvious since $g(A^*) = A^*$ for all g in G). Also it is obvious that all $g \in G$ with these properties form a subgroup of G. We define

$G^s(R_j^*/R) = G^s(M_j^*/M) = G^s(P_j^*/P) =$ (the splitting group of R_j^* over R or M_j^* over M or P_j^* over P) = (the set of all g in G for which $g(R_j^*) = R_j^*$).

We further define

$F_j^s(R_j^*/R) = F_j^s(M_j^*/M) = F_j^s(P_j^*/P) =$ (the splitting field of R_j^* over R or M_j^* over M or P_j^* over P) = (the fixed field of $G_j^s(R_j^*/R)$).

The following proposition explains the term splitting field. Let $G_j^s = G^s(R_j^*/R)$, $K_j^s = F_j^s(R_j^*/R)$, $R_j^s = R_j^* \cap K_j^s$, and $M_j^s = M_j^* \cap K_j^s$.

PROPOSITION 1.46. K_j^s is the smallest field K' between K and K^* for which R_j^* is the only local ring in K^* lying above $R_j^* \cap K'$.

PROOF. Let T be the integral closure of R in K^*. Let $T_j^s = T \cap K_j^s$, $N_i = M_i^* \cap T$, $N_i^s = N_i \cap T_j^s$, for $i = 1, \ldots, u$.

By 1.3, there exists a in T such that $a \equiv 0 \bmod N_j$ and $1 \bmod N_i$ for all $i \neq j$. Then

$$N_{K^*/K_j^s}(a) \in N_j \cap K_j^s$$

and

$$N_{K^*/K_j^s}(a) \equiv 1 \bmod N_i \quad \text{for all } i \neq j \ .$$

Therefore for all $i \neq j$, $N_j^s \not\subset N_i$, i.e., N_i does not lie above N_j^s. Hence it is enough to show that if K' is as stated, then $K_j^s \subset K'$. Now

$$g \in G(K^*/K') \Longrightarrow [g(N_j) = N_i \text{ with } i \neq j \Longrightarrow N_j \cap K' = g(N_j \cap K') = N_i \cap K' \text{ which is a contradiction}] \Longrightarrow g \in G_j^s = G(K^*/K_j^s).$$

Therefore $G(K^*/K') \subset G(K^*/K_j^s)$ and hence $K_j^s \subset K'$.

THEOREM 1.47. We have (1) $R_j^s/M_j^s = R/M$ and (2) $MR_j^s = M_j^s$ (so that R_j^s is unramified over R).

PROOF. Let T be the integral closure of R in K^*. Let $T_j^s = T \cap K_j^s$, $N_i = M_i^* \cap T$, $N_i^s = M_i^* \cap T_j^s$ for $i = 1, \ldots, u$. ... Given $a \in T_j^s$, by 1.46 and 1.3 there exists b in T_j^s such that $b \equiv a \bmod N_j^s$ and $1 \bmod N_i^s$ for all $i \neq j$, so that $b \equiv a \bmod N_j$ and $1 \bmod N_i$ for all $i \neq j$. Let $G_j^s = G_1, G_2, \ldots, G_q$ be the left cosets of G_j^s in G. Fix g_i in G_i. Then $g_1(b), \ldots, g_q(b)$ are the K_j^s/K conjugates of b (counted properly) and hence

$$c = N_{K_j^s/K}(b) = \prod_{t=1}^{q} g_i(b) \ .$$

Now $g_1(b) = b \equiv a \bmod N_j$ and for any $t \neq 1$ there exists $i \neq j$ such that $g_t(N_i) = N_j$ so that $g_t(b) \equiv 1 \bmod N_j$. Thus $g_1(b) \equiv a \bmod N_j$ and for all $t \neq 1$, $g_t(b) \equiv 1 \bmod N_j$. Hence $c \in R$ and $c \equiv a \bmod N_j$ so that $c \equiv a \bmod N_j^s$. $R_j^s/M_j^s = T_j^s/N_j^s = R/M$. This proves (1).

To prove (2), let $X_1 = N_j^s$ and let $X_2, \ldots, X_v$ be the other maximal ideals in T_j^s, so that given $i \neq j$, $N_i^s = X_t$ for some $t \neq 1$

and given $t \neq 1$, $X_t = N_i^s$ for some $i \neq j$. Then $MT_j^s = Y_1 \cap \dots \cap Y_v = Y_1 \dots Y_v$ where Y_t is primary for X_t. Let $Z = X_1 \cap \dots \cap X_v = X_1 \dots X_v$. Observe that $[a \in X_1, \notin X_t$ for $i = 2, \dots, v] \Longrightarrow$ [there exists $a^* \in T_j^s, \notin X_1$ such that $aa^* \in M \subset Y_1$, for instance let $a^* = \Pi_{i=2}^q g_i(a)] \Longrightarrow [a \in Y_1]$. By 1.3 there exists $e \in T_j^s$ such that $e \equiv 0 \bmod X_1$ and $1 \bmod X_t$ for all $t \neq 1$. Then by the above observation, $e \in Y_1$. Now $[f \in Z] \Longrightarrow [f + e \in X_1, \notin X_t$ for all $t \neq 1] \Longrightarrow [f + e \in Y_1] \Longrightarrow [f \in Y_1]$. Therefore $Z \subset Y_1$ and hence $X_1 \subset Y_1$ so that $X_1 = Y_1$. Therefore $MR_j^s = X_1R_j^s = MR_j^s$.

This completes the proof of 1.47.

Now let

I_j = the set of g in $G(K^*/K)$ for which $gu \equiv u \bmod P_j^*$ for all $u \in A^*$;

I_j^* = the set of g in $G^s(R_j^*/R)$ for which $gu \equiv u \bmod M_j^*$ for all $u \in R_j^*$.

Then clearly I_j and I_j^* are subgroups of G and G_j^s respectively; in fact, it is obvious that $g \in I_j \Longrightarrow g(M_j^*) = M_j^* \Longrightarrow g \in G_j^s$; thus I_j and I_j^* are both subgroups of G_j^s. Now let $g \in I_j$ and $v \in R_j^*$. Then $v = v_1/v_2$ with $v_1, v_2 \in A^*$ and $v_2 \notin P_j^*$. Hence

$$g(v) - v = \frac{g(v_1)}{g(v_2)} - \frac{v_1}{v_2} = \frac{g(v_1)v_2 - v_1g(v_2)}{g(v_2)v_2}$$

$$= \frac{g(v_1)v_2 - v_1v_2 + v_1v_2 - v_1g(v_2)}{g(v_2)v_2}$$

$$= \frac{v_2(g(v_1) - v_1)) - v_1(g(v_2) - v_2))}{g(v_2)v_2}$$

$$\in M_j^* \quad [\text{since } v_2 \notin M_j^* \text{ and } g \in G_j^s \Longrightarrow g(v_2) \notin M_j^*].$$

Therefore $g \in I_j \Longrightarrow g \in I_j^*$. That $g \in I_j^* \Longrightarrow g \in I_j$ is obvious. Therefore $I_j = I_j^*$.

We define

$G^i(R_j^*/R) = G^i(M_j^*/M) = G^i(P_j^*/P) =$ (<u>the inertia group</u> of R_j^* over R or M_j^* over M or P_j^* over P)

$= I_j = I_j^*$.

$F^i(R_j^*/R) = F^i(M_j^*/M) = F^i(P_j^*/P) =$ (the inertia field of R_j^* over R or M_j^* over M or P_j^* over P) = the fixed field of $G^i(R_j^*/R)$.

Let $G_j^i = G^i(R_j^*/R)$, $K_j^i = F^i(R_j^*/R)$, $R_j^s = K_j^s \cap R_j^*$, $M_j^s = K_j^s \cap M_j^*$, $R_j^i = K_j^i \cap R_j^*$ and $M_j^i = K_j^i \cap M_j^*$.

Observe that since R_j^* is the only local ring in K^* lying above R_j^s, we have: R_j^* = the integral closure of R_j^s in K^*, $G^s(R_j^*/R_j^s) = G(K^*/K_j^s)$, $F^s(R_j^*/R_j^s) = K_j^s$ and $G^i(R_j^*/R) = G^i(R_j^*/R_j^s)$.

For $a \in R_j^*$, by $\bar{a}$ we shall denote the residue class of a mod M_j^*. Let

$$H_j^* = R_j^*/M_j^*, \quad H_j^i = R_j^i/M_j^i, \quad H_j^s = R_j^s/M_j^s = R/M = H \ .$$

For $g \in G^s(R_j^*/R)$ let us define a transformation $f(g)$ of H_j^* into itself by the equation: $f(g)\bar{a} = \overline{g(a)}$. Then

(1) $f(g)$ is <u>single-valued</u>: PROOF. $\bar{a} = \bar{b} \Longrightarrow a - b \in M_j^*$ $\Longrightarrow g(a) - g(b) = g(a - b) \in M_j^*$ [since $g \in G^s(R_j^*/R)$] $\Longrightarrow \overline{g(a)} = \overline{g(b)}$.

(2) $f(g)$ <u>is an automorphism of</u> H_j^*/H. Obvious, since g is an automorphism of R_j^*/R.

THEOREM 1.48. H_j^*/H is normal and f is a homomorphism of $G^s(R_j^*/R)$ onto $G(H_j^*/H)$ and $G^i(R_j^*/R)$ = kernel of f, K_j^i/K_j^s is galois. H_j^i is the separable algebraic closure of H in H_j^* and f canonically induces an isomorphism of $G(K_j^i/K_j^s)$ onto $G(H_j^*/H) = G(H_j^i/H)$. Also we have $M_j^i = M_j^s R_j^i = M R_j^i$ (so that R_j^i is unramified over R as well as over R_j^s).

PROOF. Since all the groups in question are subgroups of $G^s(R_j^*/R)$ and all the subfields of K^* in question contain K_j^s and since $H_j^s = H$, it is clear that in the proof we may replace K by K_j^s, or in other words we may assume that (R_j^*, M_j^*) is the only local ring in K^* lying above R. Hence we may drop the subscript j all the way through and observe that now $G^s(R^*/R) = G(K^*/K)$ and $K^s = K$.

For $\bar{a} \in H^*$ let

$\bar{q}(X)$ = the minimal monic polynomial of $\bar{a}/H$; and
$q(X)$ = the minimal monic polynomial of a/K.

$a \in R^* \Longrightarrow a/R$ integral $\Longrightarrow q(X) \in R[X]$. Let $q^*(X)$ be the (monic) polynomial gotten by reducing the coefficients of $q(X)$ modulo M^*. Then $q(a) = 0 \Longrightarrow q^*(\bar{a}) = 0 \Longrightarrow \bar{q}(X)$ divides $q^*(X)$. Now K^*/K is galois

$$\Longrightarrow q(X) = \prod_{t=1}^{\deg q} (X - a_t)$$

with $a_t \in K^*$. Now $q(X) \in R[X] \subset R^*[X]$ and hence by 1.16 $a_t \in R^*$ for all t. Therefore

$$\bar{q}(X) = \prod_{u=1}^{\deg \bar{q}} (X - \bar{a}_{t_u}), \quad \text{with} \quad \bar{a}_{t_u} \in H^* .$$

Therefore H^*/H is normal. It is obvious that f is a homomorphism into. Let H' = the separable algebraic closure of H in H^*. Then by 1.42, H'/H is finite algebraic and hence a galois extension. Assume that $\bar{a}$ is a primitive element of H'/H. Then there exists g_u in $G(K^*/K)$ such that $g_u(a) = a_{t_u}$. Then $f(g_u)\bar{a} = \bar{a}_{t_u}$. Therefore f is onto.

For $g \in G(K^*/K)$, $[f(g) = 1] \Longleftrightarrow$ [for all b in R^* we have $f(g)\bar{b} = \bar{b}$, i.e., $\overline{g(b)} = \bar{b}$, i.e., $g(b) \equiv b \bmod M^*$] $\Longleftrightarrow [g \in G^i(R^*/R)]$. Therefore kernel $f = G^i(R^*/R) = G(K^*/K^i)$. Therefore $G(K^*/K^i)$ is a normal subgroup of $G(K^*/K)$. Therefore K^i/K is galois.

Also $[K^i : K]$ = [index of $G(K^*/K^i)$ in $G(K^*/K)$] = [order of $G(H^*/H)$] = $[H' : H]$.

Obviously $G^i(M^*/M^i) = G^i(M^*/M) = G(K^*/K^i)$. Therefore $G(H^*/H^i)$ is isomorphic to $G(K^*/K^i)/G(K^*/K^i) = 1$. Therefore H^*/H^i is purely inseparable. Hence $H' \subset H^i$. Therefore $[H^i : H]_s = [H' : H] = [K^i : K]$. Hence by 1.41, $H^i = H'$ and $MR^i = M^i$.

Let q denote the canonical isomorphism of $G(H^*/H)$ onto $G(H^i/H)$. Let $F = qf$. Then

$$\boxed{\begin{array}{c} F : G(K^*/K) \xrightarrow{\text{homo onto}} G(H^i/H) \\ \text{kernel } F = G(K^*/K^i) \end{array}} \; .$$

For $g \in G(K^*/K)$ set $p(g) = g/K^i$. Then

$$\boxed{\begin{array}{c} p : G(K^*/K) \xrightarrow{\text{homo onto}} G(K^i/K) \\ \text{kernel } p = G(K^*/K^i) \end{array}} \; .$$

Let $f^* = Fp^{-1}$. Then f^* is an isomorphism of $G(K^i/K)$ onto $G(H^i/H)$.

As f was defined for R^*/R we define $f^{(i)}$ for R^i/R. Then for $g \in G(K^*/K)$ we have $f^{(i)}(p(g)) = f$. Therefore $g = f^*$. This completes the proof.

PROPOSITION 1.49. Let R be a normal local domain with quotient field K, K^*/K a galois extension, R^* a local ring in K^* lying above R, K' a field between K and K^* and $R' = K' \cap R^*$. Then

$$G^s(R^*/R') = G^s(R^*/R) \cap G(K^*/K'),$$
$$G^i(R^*/R') = G^i(R^*/R) \cap G(K^*/K'),$$
$$F^s(R^*/R') = \text{compositum of } F^s(R^*/R) \text{ and } K',$$
$$F^i(R^*/R') = \text{compositum of } F^i(R^*/R) \text{ and } K'.$$

PROOF. Straightforward application of galois theory.

PROPOSITION 1.50. Let (R, M) and (S, N) be normal local domains with a common quotient field K; let K^* be a galois extension of K and let (R^*, M^*), (S^*, N^*) be local rings in K^* lying above R and S respectively. Assume that $R \subset S$ and $R^* \subset S^*$. Then

(1) $G^i(S^*/S) \subset G^i(R^*/R)$, and

(2) If S has center M in R, then $G^s(S^*/S) \subset G^s(R^*/R)$.

PROOF. Let R' and S' be the integral closures of R and S in K^* respectively. Then $R' \subset S'$. Let $N' = S' \cap N^*$ and $M' = R' \cap M^*$. Let $P' = R' \cap N'$. Then $N^* \neq S^* \Longrightarrow 1 \notin N^* \Longrightarrow N^* \cap R^* \neq R^* \Longrightarrow$ $N^* \cap R^* \subset M^* \Longrightarrow P' = (N^* \cap R^*) \cap R' \subset M^* \cap R' = M'$. Now $g \in G^i(S^*/S) \Longrightarrow$ $[a \in R' \Longrightarrow a - g(a) \in P' \subset M'] \Longrightarrow g \in G^i(R^*/R)$. Now assume that $N \cap R = M$. Then $P' \cap R = (R' \cap N') \cap R = N' \cap R = (N^* \cap S') \cap R =$ $N^* \cap R = (N^* \cap S) \cap R = N \cap R = M$, and hence $P' = M'$. Therefore $g \in G^s(S^*/S) \Longrightarrow [a \in M' \Longrightarrow a \in N' \Longrightarrow g(a) \in N' \Longrightarrow g(a) \in N' \cap R' =$ $P' = M'] \Longrightarrow g \in G^s(R^*/R)$.

CHAPTER II: VALUATION THEORY

8. Ordered abelian groups. In this section Z, Q and R will stand for the additive groups of integers, rational numbers and real numbers, respectively. In this section we shall consider ordered abelian groups a little more thoroughly than necessary, in order to give the reader a feeling about these groups.

A simply ordered set S is a collection of objects a, b, ..., with a binary relation $\geq$ with the following properties:

(1) $a \geq a$ for each a (reflexive)

(2) $a \geq b$ and $b \geq c \Longrightarrow a \geq c$ (transitive)

(3) $a \geq b$ and $b \geq a \Longrightarrow a = b$ ⎫ (trichotomy)

(4) $a, b \in S \Longrightarrow a \geq b$ or $b \geq a$ ⎭

We write $a > b$ if $a \geq b$ and $a \neq b$; also we write $b < a$ if $a > b$.

An ordered abelian group S is an abelian group which is simply ordered such that $a, b \in S$ with $a \geq b \Longrightarrow a + c \geq b + c$ for all $c \in S$ (i.e., the group operation is compatible with the ordering or vice versa). When talking about ordered abelian groups, by an ordered isomorphism (or homomorphism) is meant an isomorphism (respectively homomorphism) which preserves order. Let $a \in S$. a is said to be positive or negative according as $a > 0$ or $a < 0$ respectively. Observe that $a > 0 \Longleftrightarrow -a < 0$. We define

$$|a| = \begin{cases} a & \text{if } a \geq 0 \\ -a & \text{if } a < 0 \end{cases} .$$

For a subset T of S, by T^+ we denote the positive part of T, i.e., the set of positive elements in T; the complement of T^+ in S^+ will be called the positive complement of T (in S). Observe that S^+ is a subsemigroup of S such that $a \in S \Longrightarrow$ one and only one of the following three conditions holds: (1) $a \in S^+$, (2) $-a \in S^+$, (3) $a = 0$. Conversely, given an abelian group S and a subsemigroup S^+ with the above property, S can be (uniquely) converted into an ordered abelian group by setting $a > b \Longleftrightarrow a - b \in S^+$.

Examples of ordered abelian groups. Z, Q, R with the usual ordering and any subgroup of an ordered abelian group (in particular, of Q or R) with the induced ordering are all ordered abelian groups. Now let $S_1, \ldots, S_p$ be a finite number of ordered abelian groups and consider their direct sum $S_1 \oplus \cdots \oplus S_p$, the elements of which are of the form $(s_1, \ldots, s_p)$ with $s_i \in S_i$; let $(t_1, \ldots, t_p)$ with $t_i \in S_i$ be another element of this direct sum different from $(s_1, \ldots, s_p)$, then we can choose q such that $s_i = t_i$ for all $i < q$ and $s_q \neq t_q$. We set $(s_1, \ldots, s_p) > (t_1, \ldots, t_p)$ or $(t_1, \ldots, t_p) < (s_1, \ldots, s_p)$, according as $s_q > t_q$ or $t_q < s_q$. This converts the direct sum into an ordered abelian group which we shall call the lexicographically (i.e., as in a dictionary) ordered direct sum of $S_1, \ldots, S_p$, and we shall denote it by $S_1 \otimes \cdots \otimes S_p$. If $S_1 = \cdots = S_p = S$ then we shall write S^n for $S_1 \otimes \cdots \otimes S_n$. It will be seen that R^n and its subgroups are enough for algebro-geometric purposes to serve as value groups of valuations.

Now let S be an ordered abelian group. A nonempty subset T of S will be called a segment if $t \in T, s \in S$ with $|s| \leq t \Longrightarrow s \in T$. Let $T_1 \neq T_2$ be two segments; then either there exists $t_1 \in T_1, \notin T_2$ or $t_2 \in T_2, \notin T_1$; if $t_1 \in T_1, \notin T_2$ then $t \in T_2 \Longrightarrow [|t| > |t_1| \Longrightarrow t_1 \in T_2$ which is a contradiction$] \Longrightarrow |t| \leq |t_1| \Longrightarrow t \in T_1$, i.e., $T_2 \subset T_1$; similarly, if $t_2 \in T_2, \notin T_1$ then $T_1 \subset T_2$. Thus the set of all the segments of S is simply ordered by inclusion. An isolated subgroup of S is a subgroup which is a segment; it can easily be shown that a segment T is an isolated subgroup if and only if any one of the following four conditions is satisfied:

(1) T^+ is a semigroup;
(2) $t \in T, n \in Z \Longrightarrow nt \in T$;
(3) $t \in T, n \in Z^+ \Longrightarrow nt \in T$;
(4) $t \in T^+, n \in Z^+ \Longrightarrow nt \in T^+$.

Since the set of all the segments is simply ordered by inclusion, hence so is the set of all the isolated subgroups of S; the order type of the set of the nonzero isolated subgroups of S is called the rank of S and is denoted by $\rho(S)$. If there are only a finite number n of isolated subgroups, then $\rho(S)$ is the order type of the sequence $1, 2, \ldots, n$ and we set $\rho(S) = n$. A subset T^* of S is called an upper segment if $T^* \subset S^+$ and if $t \in T^*$, $s \in S$ with $s > t \Longrightarrow s \in T^*$. An upper segment T^* is said to be isolated if $s, t \in S^+, \notin T^* \Longrightarrow s + t \notin T^*$. Observe that a subset T^* of S is an upper segment if and only if it is the positive complement of a segment T and that T^* is an isolated upper segment if and only if it is the positive complement of an isolated subgroup. Hence there is a one-to-one inclusion reversing correspondence

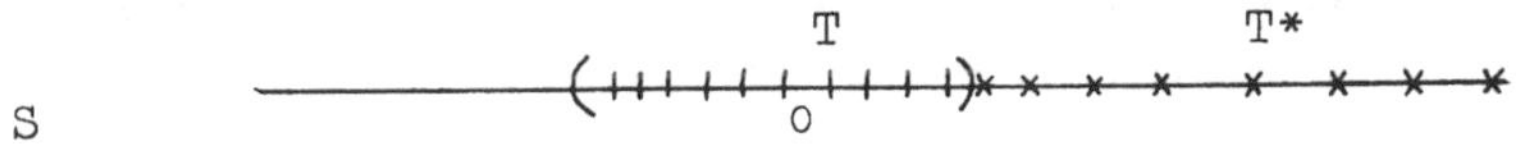

between the set of all the segments and the set of all the upper segments and it induces a one-to-one inclusion reversing correspondence between the set of all the isolated subgroups and the set of all the isolated upper segments, and hence in particular the set of all the upper segments and the set of all the isolated upper segments are simply ordered by inclusion.

Let h and k be two order types (an order type can be defined to be an equivalence class of simply ordered sets, the equivalence being the existence of an order preserving one-to-one map) and let H and K be two disjoint simply ordered sets of order types h and k respectively. Let $L = H \cup K$ and order L as follows: Given $a, b \in L$; if $a, b \in H$ then $a \geqq b$ if and only if $a \geqq b$ in H; if $a, b \in K$ then $a \geqq b$ if and only if $a \geqq b$ in K; if $a \in H$ and $b \in K$, then $a < b$. Then L is simply ordered and the order type ℓ of L depends only on h and k and we set $\ell = h + k$. Observe that if H and K are both finite, say consisting of H^* and K^* elements, respectively, then H, K and L are order isomorphic to the sequences $1, 2, \ldots, H^*$; $1, 2, \ldots, K^*$; and $1, 2, \ldots H^* + K^*$, respectively; so that in this case the equation $\ell = h + k$ can be interpreted as giving the ordinary sum of two integers.

EXERCISE. Is $h + k = k + h$ always true?

Now let S and S^* be ordered abelian groups and let f be an order homomorphism of S into S^*. Let $s, t \in S$. Then $|s| \leqq |t| \Longrightarrow |f(s)| \leqq |f(t)|$ and hence $|s| \leqq |t|$ and $f(t) = 0 \Longrightarrow |f(s)| = 0 \Longrightarrow f(s) = 0$. Therefore $f^{-1}(0)$ is an isolated subgroup of S. Now let $f^{-1}(0) = T$ and replacing S^* by $f(S)$ assume that f is onto. Then we have $\rho(S) = \rho(T) + \rho(S^*)$. [PROOF. It is enough to show that f induces a one-to-one correspondence between those isolated subgroups of S which properly contain T and the isolated subgroups of S^*, i.e., to show that if H is a subgroup of S properly containing T then H is isolated if and only if $f(H)$ is isolated. This follows from the fact that $a, b \in S$, $a \geqq b \Longrightarrow f(a) \geqq f(b)$.]

Now, conversely, let S^* be an isolated subgroup of an ordered abelian group S. Given $A \neq B$ in S/T, fix $a, b \in S$ such that $a \in A$ and $b \in B$. Set $A > B$ or $B > A$ according as $a > b$ or $b > a$. This turns S/T into an ordered abelian group. [PROOF. The only thing to be verified is that $>$ is well defined. Say $a > b$ and let $a', b' \in S$ such that $a' \in A$ and $b' \in B$. Then $a' = a + t_1$ and $b' = b + t_2$ with $t_1, t_2 \in T$. $a' \leqq b' \Longrightarrow [t_1 < t_2$ (since $a > b$)$] \Longrightarrow [a - b = a' - t_1 + t_2 - b' = (a' - b') + (t_2 - t_1) \leqq t_2 - t_1$, $a - b > 0$, $t_2 - t_1 > 0$, $t_2 - t_1 \in T$, T is isolated$] \Longrightarrow [a - b \in T] \Longrightarrow A = B$

which is a contradiction. Therefore $a' > b'$] and we shall denote it by $S//T$ (if it is clear from the context, we shall write S/T for $S//T$).

Observe that if $S_1, \ldots, S_p$ are ordered abelian groups and $0 < m < p$, then $S_{m+1} \otimes \cdots \otimes S_p$ [$= \underbrace{0 \otimes \cdots \cdots}_{(m \text{ times})} \otimes (0) \otimes S_{m+1} \otimes \cdots \otimes S_p$] is an isolated subgroup of $S_1 \otimes \cdots \otimes S_p$ and $S_1 \otimes \cdots \otimes S_p//S_{m+1} \otimes \cdots \otimes S_p$ is canonically order isomorphic with $S_1 \otimes \cdots \otimes S_m$ so that $\rho(S_1 \otimes \cdots \otimes S_p) = \rho(S_{m+1} \otimes \cdots \otimes S_p) + \rho(S_1 \otimes \cdots \otimes S_m)$ and hence by induction: $\rho(S_1 \otimes \cdots \otimes S_p) = \rho(S_p) + \rho(S_{p-1}) + \cdots + \rho(S_1)$.

EXERCISE. Let T be an isolated subgroup of an ordered abelian group S. Is S order isomorphic to $(S//T) \otimes T$? Is S group isomorphic to $(S/T) \oplus T$?

Now let S be a torsion free (i.e., S has no nonzero elements of finite order) abelian group. If for $s \in S$ and integers m and n with $n \neq 0$ there exists s' in S such that $ns' = ms$, then s' is unique and $n^*s' = m^*s'$ whenever $m/n = m^*/n^*$ and hence we may denote s' by $(m/n)s$. Let $B = \{b_j\}$, $j \in J$, be a family of elements b_j of S. An element s of S is said to depend rationally on B if there exist $m_1, \ldots, m_q \in Z$, $0 \neq n \in Z$ and $j_1, \ldots, j_q \in J$ such that $ns = m_1 b_{j_1} + \cdots + m_q b_{j_q}$ (e.g., if there exist $n_1, \ldots, n_q \in Q$ and $j_1, \ldots, j_q \in J$ such that $n_i b_{j_i}$ exists and $s = n_1 b_{j_1} + \cdots + n_q b_{j_q}$) then s is said to depend rationally on B. If some b_j depends rationally on the remaining b_j's, then B is said to be <u>rationally dependent</u>, otherwise B is <u>rationally independent</u>. If B is rationally independent, and if every element of S depends rationally on B, then B is said to be a <u>rational basis</u> of S. There exist rational bases of S and the cardinality of any two rational bases is the same. [PROOF: (The proof is entirely analogous to the corresponding theorems for the various kinds of bases, e.g., transcendence bases of a field, vector space basis of a vector space -- indeed, S can be considered to be a vector space or rather a module over Z.) Let W be the set of all the rationally independent subsets of S partially ordered by inclusion, then W has the Zorn property, etc.] and this cardinality is called the <u>rational rank</u> of S and is denoted by $r(S)$. If for every $s \in S$ and $0 \neq n \in Z$ there exists s' in S with $ns' = s$, then S is said to be <u>rationally complete</u>.

LEMMA 2.1. Let S be a torsion free abelian group. Then there exists a rationally complete abelian overgroup S^* of S such that $s^* \in S^* \Longrightarrow$ there exists $0 \neq n \in Z$ such that $ns^* \in S$. If S' is any other such group, then there exists a unique S-isomorphism of S^* onto S'. (We call S^* a <u>rational completion</u> of S.)

PROOF. Consider the set of triples (m, n, s) with $m \in Z$, $0 \neq n \in Z$ and $s \in S$. Identify (m, n, s) with (m', n', s') if and only if $n'ms = nm's'$ and then identify $(1, 1, s)$ with s; this gives S^* (just like getting Q from Z).

LEMMA 2.2. Let S be a torsionfree abelian group and S^* a rational completion of S. Then $r(S) = r(S^*)$ and every rational basis of S is a rational basis of S^*. If S is ordered, then the ordering of S can be uniquely extended to S^* (when this is done, S^* is called an <u>ordered rational completion</u> of S) and $\rho(S) = \rho(S^*)$.

PROOF. Straightforward.

LEMMA 2.3. An ordered abelian group S is torsion free.

PROOF. Let $0 \neq s \in S$ and $0 \neq n \in Z^+$ with $ns = 0$. Replacing s by $-s$, if necessary, we may assume that $s > 0$. Then $s > 0 \Longrightarrow ns = s + \dots + s > 0$ which is a contradiction.

DEFINITION 2.4. An abelian group S will be called an <u>integral direct sum</u> (or finitely generated free abelian) if there exists a finite number of elements $s_1, \dots, s_n$ of S such that any element s of S can uniquely be expressed as $s = m_1 s_1 + \dots + m_n s_n$ with $m_i \in Z$; if this is so, then S is torsionfree and $r(S) = n$; $(s_1, \dots, s_n)$ is called an <u>integral basis</u> of S.

LEMMA 2.5. Let S be an ordered abelian group. Then S is an integral direct sum $\Longleftrightarrow \rho(S) = n < \infty$ and if $0 = S_0 < S_1 < \dots < S_n = S$ is the sequence of isolated subgroups of S, then S_i/S_{i-1} is an integral direct sum for $i = 1, \dots, n$ and $S = T_1 \otimes \dots \otimes T_n$ with $T_{n-i} = S_{i+1}/S_i$ [hence integral bases of $T_1, \dots, T_n$ put together give an integral basis of S].

PROOF. ($\Longleftarrow$) Obvious. ($\Longrightarrow$) Let S be free with $r(S) = m$. Let S_1 be the smallest (nonzero) isolated subgroup of S. It is well known that a subgroup of a free abelian group with a finite number of generators is again such. Let $(f_1, \dots, f_n)$ and $(g_1, \dots, g_q)$ be integral bases of S and S_1, respectively. It is well known (see [W2], section 108 or Seifert-Threllfall: <u>Topologie</u>, Chapter XII, section 86, or Eilenberg-Steenrod: <u>Foundations of Algebraic Topology</u>, Chapter V, section 7) that we may arrange matters so as to have $g_i = m_i f_i$ with $0 \neq m_i \in Z^+$ (m_i divides m_{i+1}) for $i = 1, \dots, q$. Then $g_i \in S_1 \Longrightarrow f_i \in S_1 \Longrightarrow m_i = 1$ for $i = 1, \dots, q$. Thus $S = T \otimes S_1$, where T is the subgroup of S generated by $f_{q+1}, f_{q+2}, \dots, f_n$. Now apply induction to n.

DEFINITION 2.6. Let S be an ordered abelian group. S is said to be <u>archimedian</u> if $\rho(S) = 1$. S is said to be <u>discrete archimedian</u> if S is archimedian and if S does not contain any strictly descending infinite sequence of positive elements. S is said to be discrete if $\rho(S) = n < \infty$ and if $0 = S_0 < S_1 < \cdots < S_n = S$ is the sequence of isolated subgroups of S, then S_i/S_{i-1} is discrete (archimedian) for $i = 1, \ldots, n$.

LEMMA 2.7. An ordered abelian group S is archimedian $\Longleftrightarrow$ [$0 \neq s \in S$ and $t \in S \Longrightarrow$ there exists n in Z such that $ns > t$].

PROOF. ($\Longrightarrow$) Otherwise there exist $s, t \in S$, $s \neq 0$ such that $ns < t$ for all $n \in Z$. Let H be the set of all elements h in S for which $|nh| < t$ for all $n \in Z$. Then H is an isolated subgroup of S with $0 \neq H \neq S$ so that $\rho(S) > 1$ which is a contradiction.

($\Longleftarrow$) Otherwise there exists an isolated subgroup H of S with $0 \neq H \neq S$. Let $s > 0$ in S^+ with $s \notin H$ and let $0 \neq h \in H$. Then for all n in Z, $nh \in H$ so that $nh < s$ which is a contradiction.

LEMMA 2.8. An archimedian ordered abelian group S is discrete $\Longleftrightarrow$ S is order isomorphic to Z.

PROOF. ($\Longleftarrow$) Obvious. ($\Longrightarrow$) The nonexistence of a smallest positive element in S would imply the existence of a strictly descending infinite sequence of positive elements which would be a contradiction. Therefore there exists a smallest positive element e in S. Now $s \in S$ and $s \neq ne$ for all $n \in Z \Longrightarrow |s| \neq ne$ for all $n \in Z \Longrightarrow$ (by 2.7) there exists $n \in Z$ such that $(n - 1)e < |s| < ne$ so that $0 < ne - |s| < e$ which is a contradiction. Therefore every element of S is of the form ne with $n \in Z$. Since e is of infinite order $ne \longrightarrow n$ gives the (order) isomorphism of S with Z.

LEMMA 2.9. An ordered abelian group S is archimedian $\Longleftrightarrow$ S is order isomorphic to a subgroup of R.

PROOF. ($\Longleftarrow$) Obvious. ($\Longrightarrow$) Let us define a map f of S into R as follows: $f(0) = 0$; fix $s > 0$ in S and let $f(s) = 1$. Given $t \in S$, let $A_t = \{m/n \in Q \mid ms \leq nt\}$ and $B_t = \{m/n \in Q \mid ms > nt\}$. Then A_t, B_t is a dedekind cut. Let $f(t)$ = the real number defined by this cut. Now it can be shown that f is an order isomorphism.

PROPOSITION 2.10. Let S be an ordered abelian group with $\rho(S) = n < \infty$. Then S is order isomorphic to a subgroup of R^n.

If S is rationally complete, then we have the stronger conclusion: there exist subgroups S_i of R such that S is order isomorphic to $S_1 \otimes \cdots \otimes S_n$.

PROOF. In view of 2.2, it is enough to prove the second assertion. We make induction on n, the case $n = 1$ is covered by 2.9, so let $n > 1$ and assume true for $n - 1$. Let H be the maximal isolated subgroup of S other than S. Then H and S/H are rationally complete and of ranks $n - 1$ and 1 respectively. By the induction hypothesis, $H = S_2 \otimes \cdots \otimes S_n$ with $\rho(S_j) = 1$. Let B be a rational basis of S/H and for b in B fix b^* in S in the residue class b. Let $B^* = \{b^*\}$. Then B^* is rationally independent. Let S_1 be the set of all elements of S rationally dependent on B^*. Then S_1 is a rationally complete subgroup of S and hence under the canonical order isomorphism of S onto $S//H$, S_1 gets mapped onto $S//H$. Therefore $S = S_1 \otimes H = S_1 \otimes \cdots \otimes S_n$.

EXERCISE. In Proposition 2.10, is the stronger conclusion true without the restriction of rational completeness?

LEMMA 2.11. Let S be an ordered abelian group. Then the following two statements are equivalent:

(1) $r(S) = r < \infty$.

(2) $\rho(S) = n < \infty$ and if $0 = S_0 < S_1 < \cdots < S_n = S$ is the sequence of isolated subgroups of S, then $r(S_i/S_{i-1}) = r_i < \infty$ for $i = 1, \ldots, n$.

When (1) and hence (2) holds, then $r = r_1 + \cdots + r_n$.

PROOF. By 2.10.

LEMMA 2.12. Let S be an ordered abelian group. Then S is discrete $\Longleftrightarrow$ S is an integral direct sum and $\rho(S) = r(S) \Longrightarrow S = Z^r$.

PROOF. Straightforward, in view of what has been proved up till now.

PROPOSITION 2.13. Let S be an ordered abelian group and T a subgroup of S of finite index. Then

(1) $\rho(S) = \rho(T)$ and $r(S) = r(T)$;

(2) S is an integral direct sum $\Longleftrightarrow$ T is an integral direct sum; and

(3) S is discrete $\Longleftrightarrow$ T is discrete.

PROOF. Straightforward.

9. <u>Valuations</u>. Let K be a field and let K^x denote the multiplicative group of the nonzero elements of K. A <u>valuation</u> v of K is a homomorphism of K^x into an ordered abelian group S such that $v(a + b) \geq \min(v(a), v(b))$ for all a, b in K^x. $v(a)$ is called the value (or rather the v-value) of a. Since $v(F^x)$ is a subgroup of S, by replacing S with $v(F^x)$, we may assume that $v(F^x) = S$; then S will be called the value group of v and will be denoted by S_v. The adjectives of S_v may then be applied to v, e.g., $\rho(v) = \rho(S_v)$, $r(v) = r(S_v)$, etc. Let k be a subfield of K. If $v(a) = 0$ for all $a \in k$, we say that v is over k or v is a valuation of K/k. Given K we

can set $v(a) = 0$ for all $a \in K^x$ and get the <u>trivial valuation</u>. Unless otherwise stated, we shall exclude this from the discussion; we may thus add in the definition that: there exists $a \in F^x$ with $v(a) \neq 0$ or equivalently, that $S_v = v(F^x)$ is nontrivial. Observe that: v is a homomorphism of F^x into $S_v \Longrightarrow$ the unit of F^x goes into the unit of S_v, i.e., $v(1) = 0$ and hence $v(-1) = 0$ [since $(-1)^2 = 1$] so that $v(-a) = v(a)$, and $v(a - b) \geqq \min(v(a), v(b))$ for all $a, b \in F^x$. Hence if $v(a) < v(b)$ then $v(a \pm b) = v(a)$. We set $v(0) = \infty$ with the convention that: for all $s \in S_v$, $s + \infty = \infty + s = \infty + \infty = \infty$ and $s < \infty$; then we need not always say "if $a \neq 0$." We define

$$R_v = (\text{the valuation ring of } v) = \{a \in K \mid v(a) \geqq 0\} \quad ,$$

$$M_v = \{a \in F \mid v(a) > 0\} \quad .$$

Then R_v is a ring and $x \in F$, $x \notin R_v \Longrightarrow v(x) < 0 \Longrightarrow x \neq 0$ and $v(1/x) > 0 \Longrightarrow 1/x \in R_v$. Now let $a \in R_v$. Then $a \notin M_v \Longleftrightarrow v(a) = 0 \Longleftrightarrow a \neq 0$ and $v(1/a) = 0 \Longleftrightarrow 1/a \in R_v$. Therefore M_v is the set of nonunits in R_v and hence by 1.5, M_v is the unique maximal ideal in R_v. The field R_v/M_v is called the residue field of v and is denoted by D_v;. if v is over k we may assume that $k \subset D_v$. The image of $a \in R_v$ under the canonical homomorphism of R_v onto D_v is called the v-residue of a; if $a \notin R_v$ we set ∞ = the v-residue of a with the convention that for all $d \in D_v$, $d + \infty = \infty + d = \infty$. Let A be a subdomain of K and P a prime ideal in A; we shall say that v has center P in A if (R_v, M_v) has center P in A, i.e., if $R_v \supset A$ and $M_v \cap A = P$. If K' is a subfield of K and v' is the restriction of the valuation v of K to K', then v' is a valuation (possibly trivial) of K' and v is an extension of v' to K.

PROPOSITION 2.14. Let R be a subdomain of a field K different from K. Then R is the valuation ring of a valuation of $K \Longleftrightarrow [a \in K, a \notin R \Longrightarrow 1/a \in R]$.

PROOF. ($\Longrightarrow$) Has been proved above. ($\Longleftarrow$) [Observe that if a valuation v of K with $R = R_v$ does exist, then $R_v - M_v$ is the kernel of v so that S_v is canonically (group) isomorphic to $K^x/(R_v - M_v)$ and $M_v/(R_v - M_v) \longrightarrow S_v^+$; to prove the existence of v, we invert this procedure.] Let I be the set of units in R. Then I is a multiplicative subgroup of F^x. Let $S = F^x/I$, and let us write the operation in S additively. For $a \in F^x$ let a' denote the residue class of a modulo I. Then $a' = b' \Longleftrightarrow b = au$ with $u \in I \Longrightarrow [a \in R \Longleftrightarrow b \in R]$ and $[a \in I \Longleftrightarrow b \in I]$. Hence if we let $P = \{a' \in S \mid a \in R, a \notin I\}$ and $N = \{a' \in S \mid a \notin R\}$ then P and N are well defined disjoint subsets of S such that $0 \notin P, \notin N$ and $(0) \cup P \cup N = S$ and for $a' \in S$,

$a' \in P \Longleftrightarrow -a' \in N$. Also $a', b' \in P \Longrightarrow a, b \in R, \notin I \Longrightarrow ab \in R, \notin I \Longrightarrow a' + b' = (ab)' \in P$. Hence if we set $P = S^+$ (i.e., $a' > b' \Longleftrightarrow a' - b' \in P$, i.e., $a' > b' \Longleftrightarrow a/b \in R, \notin I$, i.e., $a' \geqq b' \Longleftrightarrow a/b \in R$) then S becomes an ordered abelian group. So it is enough to show that $a, b \in F^x \Longrightarrow (a + b)' \geqq \min(a', b')$. Say $a' \leqq b'$. Then we have to show that $(a + b)' - a' \geqq 0$, i.e.,

$$\left(\frac{a + b}{a}\right)' \geqq 0 ,$$

i.e.,

$$\left(1 + \frac{b}{a}\right)' \geqq 0 .$$

But

$$a' \leqq b' \Longrightarrow \frac{b}{a} \in R \Longrightarrow \left(1 + \frac{b}{a}\right) \in R \Longrightarrow \left(1 + \frac{b}{a}\right)' \geqq 0 .$$

DEFINITION 2.15. Let v be a valuation of a field K and let $R = R_v$. The valuation constructed in 1.14 will be called the canonical valuation given by v or by R.

DEFINITION 2.16. Let v and w be two valuations of a field K. Then v and w will be said to be equivalent if there exists an order isomorphism f of S_w onto S_v such that for all $a \in K^x$ we have: $v(a) = f(w(a))$. This is clearly an equivalence relation.

LEMMA 2.17. Let v be a valuation of a field K and let u be canonical valuation given by v. Then there exists a unique order isomorphism f of S_u onto S_v such that $v(a) = f(u(a))$ for all $a \in K^x$.

PROOF. We shall use the notation of the proof of 2.14. (Existence) Let $f(a') = v(a)$. Then $a' = b' \Longrightarrow a/b \in I = R_v - M_v \Longrightarrow v(a/b) = 0 \Longrightarrow v(a) = v(b)$. Therefore f is well defined. Also $f(a' + b') = f((ab)') = v(ab) = v(a) + v(b) = f(a') + f(b')$, and $a' \geqq b' \Longrightarrow a/b \in R \Longrightarrow f(a') - f(b') = f(a'/b') = v(a/b) \geqq 0 \Longrightarrow f(a') \geqq f(b')$. (Uniqueness) $v(a) = f(u(a)) = f(a')$.

LEMMA 2.18. Let v and w be two valuations of a field K. Then v and w are equivalent $\Longleftrightarrow R_v = R_w \Longrightarrow$ there exists a unique order isomorphism g of S_w onto S_v such that $v(a) = g(w(a))$ for all $a \in K^x$.

PROOF. v and w are equivalent $\Longrightarrow$ an order isomorphism g of S_w onto S_v such that $v(a) = g(w(a))$ for all $a \in K^x \Longrightarrow [a \in R_v \Longleftrightarrow v(a) \geqq 0 \Longleftrightarrow g(w(a)) \geqq 0 \Longleftrightarrow$ (since g is an order isomorphism) $w(a) \geqq 0 \Longleftrightarrow a \in R_w] \Longrightarrow R_v = R_w$. The rest of the assertions follow from 2.17.

DEFINITION 2.18. From now on, unless otherwise stated, we shall not distinguish between equivalent valuations.

DEFINITION 2.19. Let R be a subdomain of a field K and let p be a homomorphism of R into a field D such that $p(R) \neq 0$. We shall say that p is a <u>place</u> of K if p cannot be extended, i.e., if $[p^*$ is a homomorphism of a subdomain S of K such that $R \subset S$ and $p^* \mid R = p] \Longrightarrow [S = R]$. Now assume that p is a place. Let $M = p^{-1}(0)$. Then M is the unique maximal ideal in R (for otherwise p could be extended to R_M) and hence $p(R)$ is a subfield of D so that replacing D by $p(R)$ we may assume that $D = p(R)$. We define

$$R_p = (\text{the valuation ring of } p) = R \ ;$$

$$M_p = M \ ;$$

and

$$D_p = (\text{the residue field of } p) = D \ .$$

Since any isomorphism of a subdomain of K can be extended to K and since a nonzero homomorphism of a field is an isomorphism we conclude that $R_p = K \Longleftrightarrow p$ is an isomorphism. An isomorphism of K is called a trivial place. Unless otherwise stated, by a place we shall mean a non-trivial place. Two places p and q of K are said to be isomorphic if there exists an isomorphism f of D_p onto D_q such that $fp = q$. It is clear that p and q are isomorphic if and only if $R_p = R_q$ and when that is so, the isomorphism f is unique. The canonical homomorphism of R_p onto R_p/M_p will be called the <u>canonical place</u> isomorphic to p.

PROPOSITION 2.20. Let R be a subdomain of a field K different from K. Then there exists a place p of K with $R = R_p \Longleftrightarrow [x \in K, \notin R \Longrightarrow 1/x \in R]$.

PROOF. ($\Longleftarrow$) By 2.14 there exists a valuation v of K with $R_v = R$. Let p be the canonical homomorphism of R_v onto $D_v = R_v/M_v$. Let S be a subdomain of K with $R \subset S$ and let p^* be a homomorphism of S into an overfield D^* of D_v such that $p^* \mid R = p$. Then $[x \in S, \notin R] \Longrightarrow [1/x \in R = R_v$ so that $1/x \in M_v] \Longrightarrow [p^*(1/x) = p(1/x) = 0] \Longrightarrow [p^*(1) = p^*(x)p^*(1/x) = 0]$ which is a contradiction. Therefore $S = R$. Hence p is a place.

($\Longrightarrow$) Let $D = D_p$ and $M = M_p$. <u>We assert that given</u> $u \in K, \notin R$ <u>there exist</u> $c_1, \ldots, c_n \in M$ <u>such that</u> $1 + c_1u + \ldots + c_nu^n = 0$. Let $A = \{f(X) \in R[X] \mid f(u) = 0\}$. Then A is a prime ideal in $R[X]$. Let $p_1 : R[X] \longrightarrow D[X]$ be the natural extension of p and let $A_1 = p_1(A)$. Suppose if possible that $A_1 \neq D[X]$. Then $B^* = p_1^{-1}(A_1) \neq R[X]$ and hence there exists a maximal ideal B in $R[X]$ with $B \supset B^*$. Since $B^* \supset A$ and $B^* \supset M$, we have $B \supset A$ and $B \supset M$. Let $q : R[X] \longrightarrow R[u]$ be the R-homomorphism with $q(X) = u$. Then kernel $q = A$ and hence

$N = q(B)$ is a prime ideal in $R[u]$ with $N \supset M$. Therefore $N \cap R = M$ and hence p can be extended to $R[u]$. This is a contradiction. Therefore $A_1 = D[X]$ and hence $1 \in A_1$, i.e., there exist $f_o, \ldots, f_n$ in R with $(f_o + f_1 X + \ldots + f_n X^n) \in A$ such that $p(f_1) = \ldots = p(f_n) = 0$ and $p(f_o) = 1$. Since $p(f_o) = 1$, f_o is a unit in R and hence $c_i = f_i/f_o \in M$ and $1 + c_1 u + \ldots + c_n u^n = 0$. This proves our assertion.

To complete the proof, assume if possible that $u \in K$ with $u \notin R$ and $1/u \notin R$. Let n be minimum such that

$$1 + c_1 u + \ldots + c_n u^n = 0, \quad \text{with } c_i \in M \quad . \tag{1}$$

Let m be minimum such that

$$1 + d_1(1/u) + \ldots + d_m(1/u)^m = 0, \quad \text{with } d_i \in M \quad .$$

Say $n \geq m$. Then

$$u^m + d_1 u^{m-1} + \ldots + d_m = 0 \quad ,$$

and hence

$$c_n u^n + c_n d_1 u^{n-1} + \ldots + c_n d_m u^{n-m} = 0 \quad ,$$

i.e.,

$$e_o + e_1 u + \ldots + e_n u^n = 0, \quad \text{with } e_o, \ldots, e_{n-1} \text{ in } M \text{ and } e_n = c_n \quad . \tag{2}$$

Subtracting equation (2) from equation (1), we obtain

$$(1 - e_o) + (c_1 - e_1)u + \ldots + (c_{n-1} - e_{n-1})u^{n-1} = 0 \quad .$$

Since $1 - e_o \notin M$ we can divide by it and get

$$1 + f_1 u + \ldots + f_{n-1} u^{n-1} = 0, \quad \text{with } f_i \in M \quad ,$$

which contradicts the minimality of n. Therefore $u \in K, u \notin R \Longrightarrow 1/u \in R$.

REMARK 2.21. Thus 2.14 and 2.20 result in a one-to-one correspondence between valuation v of K and the classes of isomorphic places p of K (each such class may be replaced by the canonical member in it) such that $R_v = R_p$. If $R_v = R_p$ we shall say that v is the valuation associated with p and set $v = v_p$. If $v = v_p$ we shall say that p is a place associated with v and set $p = p_v$; sometimes we shall let p_v uniquely denote the canonical place associated with v. Observe that v is trivial $\iff v_p$ is trivial. If A is a subdomain of K and P a prime ideal of A, we shall say that p has center P in A $\iff v_p$ has center P in A.

PROPOSITION 2.22. Let A be a subdomain of a field K other than K and let H be an ideal in A with $0 \neq H \neq A$. Then there exists a place of K such that $R_p \supset A$ and $M_p \supset H$. If H is prime, there exists a place of K having center H in A.

PROOF. By Zorn's lemma, there exists a prime ideal P in A with $P \supset H$. Let f be the canonical homomorphism of A onto $A/P = D$. Let D^* be an algebraic closure of a purely transcendental extension of D whose transcendence degree over D is equal to the cardinal number of K. Then it is clear that for a subdomain S of K containing R (there exists a homomorphism of S into an overfield of D which coincides with f on R) if and only if (there exists a homomorphism of S into D^* which coincides with f on R). Let W be the set of homomorphisms g_B of domains B with $A \subset B \subset K$ into D^* such that $g_B \mid A = f$. Set $g_B \geq g^*_{B^*} \iff B \supset B^*$ and $g_B \mid B^* = g^*_{B^*}$. Then W becomes a partially ordered set which has the Zorn property and hence W contains a maximal element $p = g_B$. Now $0 \neq H \neq A \implies 0 \neq P \neq A \implies$ there exists $0 \neq a \in P$ and $b \in A - P$ so that $p(a) = f(a) = 0$ and $p(b) = f(b) \neq 0$ $\implies$ p is neither the zero homomorphism nor an isomorphism. Hence the maximality of p implies that p is a place of K with $R_p = B \supset A$ and $M_p \cap A = P \supset H$. If H is prime, we can take $P = H$.

COROLLARY 2.23. Let p be place of a field K and K^* an extension of K. Then p can be extended to a place of K^*.

It is enough to observe that a place p^* of K^* is an extension of p $\iff v_{p^*}$ has center $M_{v_p} = M_p$ in $R_{v_p} = R_p$.

Examples of valuations:

(1) Prime spots in the field Q of rational numbers: Each $q \in Q$ has a unique expression: $q = \pm \Pi p^{q_p}$ where the product is over all the prime numbers. Set $v_p(q) = q_p$.

(2) Orders of rational functions of a complex variable (zeros or poles): $k(x)$, k the field of complex numbers. For $a \in k$, let $v_a(f(x))$ = the order of the zero (or the negative of the order of the pole) of $f(x)$ at $x = a$. In both of the above examples, all the valuations are real discrete. These are enough for algebraic number theory or

algebraic function theory of one variable, i.e., the theory of algebraic curves, i.e., the theory of (closed) Riemann surfaces -- a modernistic (meaning thereby, valuation theoretic, due to Dedekind-Weber, 1882), definition of which is the set of all valuation of the given function field.

(3) $k(x, y)$; (k is a field and x, y are transcendental over k), u = a positive irrational number. For $f(x, y) = \Sigma f_{ab}, x^a y^b \in k[x, y]$ (with $f_{ab} \neq 0$), set $v(f(x, y)) = \min (a + bu)$. For $f, g \in k[x, y]$ with $g \neq 0$, set $v(f/g) = v(f) - v(g)$. This gives a valuation v of $k(x, y)/k$ and the value group of v consists of the real numbers of the form $m + nu$ where m and n are integers.

(4) In the above notation, set $w(f(x, y)) = \min \{(a, b)\}$ where the pairs of integers (a, b) are lexicographically ordered. Another way of getting w:

$$f(x, y) = x^d F(x, y), \quad \text{with } F(0, y) \neq 0 \ ,$$
$$F(x, y) = y^e F^*(y) \ , \quad \text{with } F^*(0) \neq 0 \ .$$

Then $w(f) = (d, e)$. Then w is a valuation of $k(x, y)/k$.

(5) In the above notation, set $w^*(f) = d$. Then w^* is a valuation of $k(x, y)/k$.

(6) The valuation of $k(x, y)/k$ gotten by substituting a formal series $y = \Sigma a_i x^{b_i}$ (where $b_1 < b_2 < \cdots$ are real numbers and $a_i \in k$) and looking at the lowest degree term.

LEMMA 2.24. Let v be a valuation of a field K. Then R_v is normal.

PROOF. Let $0 \neq u \in K$ with u/R_v integral. Then $u^n + a_1 u^{n-1} + \cdots + a_n = 0$ with $a_i \in R_v$. Then $u \notin R_v \Longrightarrow v(u) < 0 \Longrightarrow v(a_i u^{n-i}) > v(u^n)$ for $i = 1, \ldots, n \Longrightarrow u(0) = v(u^n)$ which is a contradiction.

PROPOSITION 2.25. Let A be a subdomain of a field K and let B be the integral closure of A in K. Then B = the intersection of all the valuation rings of valuation of K containing A where we allow the trivial valuation.

PROOF. By 2.24, B is contained in the intersection. Hence it is enough to show that given $a \in K$, $a \notin B$ there exists a valuation v of K with $R_v \supset A$ and $a \notin R_v$. Now $a \notin B \Longrightarrow a \notin A[1/a] \Longrightarrow 1/a$ is a nonunit in $A[1/a] \Longrightarrow$ there exists a nonzero prime ideal P in $A[1/a]$ with $1/a \in P$. By 2.22, there exists a valuation v of K having center P in $A[1/a]$. Then $R_v \supset A[1/a] \supset A$ and $M_v \cap A[1/a] = P$, so that $1/a \in M_v$, i.e., $v(1/a) > 0$ so that $v(a) < 0$ and hence $a \notin R_v$.

COROLLARY 2.26. K/A algebraic $\iff K = B \iff$ there does not exist any nontrivial valuations v of K with $R_v \supset A$.

COROLLARY 2.27. Another proof of 1.16: The second assertion follows from the first. To prove the first assertion, in view of 2.25, we may assume that A is the valuation ring of a valuation v of K (if $A = K$ there is nothing to prove). For $a(X) = \Sigma a_t X^t$ and $0 \neq b(X) = \Sigma b_t X^t$ in $K[X]$ set $w(a(X)/b(X)) = \min\{v(a_t)\} - \min\{v(b_t)\}$. Then w is an extension of v to $K(X)$. $h(X) \in R_v[X] \implies w(h(X)) \geq 0$. Since $f(X)$ and $g(X)$ are monic, $w(f(X)) \leq 0$ and $w(g(X)) \leq 0$. Now $f(X) \notin R_v[X] \implies w(f(X)) < 0 \implies w(h(X)) = w(f(X)) + w(g(X)) < 0$ which is a contradiction. Hence $f(X) \in R_v[X]$ and similarly $g(X) \in R_v[X]$.

COROLLARY 2.28. Another proof of 1.28 (1): Let P^* be any prime ideal in A^* lying above P. By 2.22 there exists a valuation v of K^* having center P^* in A^*. Then v has center P in A and hence v has center M in R. Since R_v is normal, we must have $R_v \supset R^*$. Since $(M_v \cap R^*) \cap R = M$, we must have $M_v \cap R^* = M_i^*$ for some i. Therefore $P^* = M_v \cap A^* = M_i^* \cap A^* = P_i^*$.

Now let v be a valuation of a field K. Let I be a nonunit ideal in R_v. Then $a \in I$, $b \in K$ with $v(b) \geq v(a) \implies b = a\frac{b}{a}$ and $v(b/a) \geq 0$, i.e., $b/a \in R_v \implies b \in I$. Also $I \subset M_v \implies$ for all $a \in I$, $v(a) > 0$. Thus $U(I) = \{v(a) \mid a \in I\}$ is an upper segment of S_v. Conversely, given an upper segment U of S_v, let $I(U) = \{a \in R_v \mid v(a) \in U\}$. Then

$$a, b \in I(U) \implies v(a), v(b) \in U \implies v(a \pm b) \geq \min (v(a), v(b)) \quad ,$$

and hence $v(a \pm b) \in U \implies a \pm b \in I(U)$. Also

$$a \in I(U),\ c \in R_v \implies v(a) \in U$$

and

$$v(ca) = v(c) + c(a) \geq v(a) \implies ca \in I(U) \quad .$$

Also $v(1) = 0 \implies 1 \notin I(U)$. Therefore $I(U)$ is a nonunit ideal in R_v. Now

> $a \in I(U(I)) \iff v(a) \in U(I) \iff$ there exists $a^* \in I$ such that $v(a) = v(a^*)$, i.e., such that $a = a^*d$ with $d = a/a^*$ so that $v(d) = 0$, i.e., d a unit in $R_v \iff a \in I$.

Thus $I(U(I)) = I$. Similarly we can show that $U(I(U)) = U$. Therefore

$$I \longrightarrow U(I)$$
$$I(U) \longleftarrow U$$

is a one-to-one correspondence between the nonunit ideals in R_v and the upper segments in S_v and this correspondence is obviously inclusion reversing and hence we get a one-to-one inclusion preserving correspondence

$$I \longrightarrow S(I) = \{s \in S_v \mid |s| < v(a) \text{ for all } a \in I\}$$

between the nonunit ideals in R_v and the segments of S_v. Therefore the set of ideals in R_v is simply ordered by inclusion. Now the upper segment U is isolated $\Longleftrightarrow$ [$a, b \in S_v^*$ with $a \notin U$, $b \notin U \Longrightarrow a + b \notin U$] $\Longleftrightarrow$ [$a, b \in R_v$ with $a \notin I(U)$, $b \notin I(U) \Longrightarrow ab \notin I(U)$] $\Longleftrightarrow$ $I(U)$ is a prime ideal. In other words I is a prime ideal $\Longleftrightarrow$ $S(I)$ is an isolated subgroup. Observe that $S(M_v) = 0$. If we define the rank $\rho(v)$ of R_v to be the order type of the set of all nonmaximal prime ideals of R_v, then we can state our conclusions as follows:

PROPOSITION 2.29. There is a one-to-one inclusion preserving correspondence between the nonunit ideals in R_v and the segments of S_v and in this correspondence prime ideals correspond to isolated subgroups and we have $\rho(v) = \rho(S_v)$.

10. Specialization and composition of valuations.

LEMMA 2.30. Let v be a valuation of a field K and $S \neq K$ a subdomain of K containing R_v. Then S is the valuation ring of a valuation of K.

PROOF. A direct consequence of 2.14.

LEMMA 2.31. Let v and w be valuations of a field K such that $R_w \supset R_v$. Then

(1) $P = M_w \cap R_v$ is a nonzero prime ideal in R_v, $R_w = (R_v)_P$ and $M_w = P$;

(2) If w has center M_v in R_v, then $R_w = R_v$, i.e., $w = v$.

PROOF. Proof of (2). $a \in K$, $a \notin R_v \Longrightarrow v(a) < 0 \Longrightarrow v(1/a) > 0 \Longrightarrow 1/a \in M_v \Longrightarrow 1/a \in M_w \Longrightarrow w(1/a) > 0 \Longrightarrow w(a) < 0 \Longrightarrow a \notin R_w$.

Proof of (1). $P \neq 0$ since w is a (nontrivial) valuation and K is the quotient field of R_v. By 2.30, $(R_v)_P = R_u$ and $PR_u = M_u$, where u is a valuation of K. Then w has center M_u in R_u. Hence by (2), $R_w = R_u$. It remains to be shown that $M_w = P$, i.e., that $M_u = P$, i.e., that $M_u \subset P$. Now $0 \neq a \in M_u \Longrightarrow 1/a \notin R_u \Longrightarrow 1/a \notin R_v \Longrightarrow a \in R_v \Longrightarrow a \in M_u \cap R_v = P$.

DEFINITION 2.32. If v and w are valuations of a field K with $R_w \supset R_v$ we shall say that v is a specialization of w, in symbols:

$w \longrightarrow v$; if $w \longrightarrow v$, we shall also write $w_p \longrightarrow v_p$ for corresponding places, and say that v_p is a specialization of w_p. Observe that $(w \longrightarrow v$ and $v \longrightarrow w) \Longleftrightarrow (w = v)$ and $(w_p \longrightarrow v_p$ and $v_p \longrightarrow w_p) \Longleftrightarrow (w_p$ and v_p are isomorphic). More generally, if (R, M) is a local domain and (S, N) is a quotient ring of R with respect to a nonzero prime ideal, then we shall say that (R, M) (or R) is a specialization of (S, N) (or S), in symbols: $(S, N) \longrightarrow (R, M)$ (or $S \longrightarrow R$). Again, observe that $(S \longrightarrow R$ and $R \longrightarrow S) \Longleftrightarrow (S = R)$.

PROPOSITION 2.33. Let v be a valuation of a field K. Let A be the set of prime ideals in R_v. For $P \in A$ let $S(P) = (R_v)_P$. Let $A^* = \{S(P) \mid P \in A\}$. If $P \neq 0$ then $S(P)$ is the valuation ring of a valuation of K (if $P = 0$, then $S(P) = K$) and we have $PS(P) = P$. Also A^* is exactly the set of domains between R_v and K and it is simply ordered by inclusion.

PROOF. A direct consequence of 2.29, 2.30 and 2.31.

Now let w and v be distinct valuations of a field K with $w \longrightarrow v$ so that M_w is a prime ideal in R_v and $0 \neq M_w \neq M_v$. Let $\bar{S}$ be the isolated subgroup of S_v corresponding to the prime ideal M_w, let f be the canonical order homomorphism of S_v onto $S^* = S_v//\bar{S}$ and let $w^* = fv$. Since $(R_w\, mM_w)$ is the quotient ring of R_v with respect to M_w, we have that $R_w - M_w$ consists exactly of the set of elements of the form a/b with $a, b \in R_v - M_w$. Now $a, b \in R_v - M_w \Longrightarrow v(a), v(b) \in \bar{S} \Longrightarrow v(a/b) \in \bar{S} \Longrightarrow a/b \in M_w$. Thus $x \in R_w - M_w \Longrightarrow x \in v^{-1}(\bar{S})$. Conversely, $y \in v^{-1}(\bar{S}) \Longrightarrow v(y) \in \bar{S} \Longrightarrow$ [if $y \in R_v$ then $y \in R_v - M_w$ so that $y = y/1 \in R_w - M_w$; if $y \notin R_v$, then $1/y \in R_v$ and $v(1/y) \in \bar{S}$ so that $1/y \in R_v - M_w$ and hence $y = 1/(1/y) \in R_w - M_w] \Longrightarrow y \in R_w - M_w$. Thus we have shown that kernel $w = R_w - M_w = v^{-1}(\bar{S}) =$ kernel w^*. Now let $x \in K^x$. Then $w^*(x) > 0 \Longleftrightarrow v(x)$ is in the positive complement of $\bar{S}$ in $S_v \Longleftrightarrow x \in M_w$. Therefore w^* is a valuation of K equivalent to w, i.e., $w^* = w$ and we may take S^* to be the value group S_w of w.

Furthermore, let p_w be the canonical place associated with w, i.e., p_w is the canonical homomorphism

$$p_w : R_w \longrightarrow R_w/M_w = D_w \quad .$$

Let $p_w(R_v) = \bar{R}_v$, $p_w(M_v) = \bar{M}_v$, and for $a \in R_w$ let $\bar{a} = p_w(a)$. Observe that $p_w^{-1}(D_w^x) = R_w - M_w = v^{-1}(\bar{S})$. For $\bar{a} \in D_w^x$ let

$$\bar{v}(\bar{a}) = v(a) \in \bar{S} \quad .$$

Then $\bar{v}$ is a transformation of D_w^x onto $\bar{S}$.

(1) $\bar{v}$ <u>is single valued</u>: $a, b \in R_w$ with $\bar{a} = \bar{b} \neq 0 \Longrightarrow a$, $b \notin M_w$ and $a = b + d$ with $d \in M_w$ so that $v(d) > v(a)$, $v(b) \Longrightarrow v(a) = v(b)$.

(2) $\bar{v}$ <u>is a homomorphism</u>: Obvious.

(3) $\bar{v}$ <u>is a valuation of</u> D_w^x: Let $\bar{a}, \bar{b} \in D_w^x$. Then $\bar{v}(\bar{a} + \bar{b}) = v(a + b) \geq \min(v(a), v(b)) = \min(\bar{v}(\bar{a}), \bar{v}(\bar{b}))$.

(4) $R_{\bar{v}} = \bar{R}_v$ <u>and</u> $M_{\bar{v}} = \bar{M}_v$: $\bar{v}(\bar{a}) \geq 0 \Longleftrightarrow a \in R_v \Longleftrightarrow \bar{a} \in \bar{R}_v$. Also $\bar{v}(\bar{a}) > 0 \Longleftrightarrow a \in M_v \Longleftrightarrow \bar{a} \in \bar{M}_v$.

Thus, given valuations $w \longrightarrow v$ of K, with $w \neq v$, we have decomposed v into w and $\bar{v}$ (a valuation of D_w^x). We write

$$w = w \circ \bar{v} \quad .$$

Conversely, let w be a valuation of K, and $\bar{v}$ a valuation of D_w^x. Let $R = p_w^{-1}(R_{\bar{v}})$. Then R is the valuation ring of a valuation v of K and $v = w \circ \bar{v}$.

PROOF. $a \in R_w, \notin R \Longrightarrow p_w(a) \notin R_{\bar{v}} \Longrightarrow a \notin M_w$ so that $1/a \in R_w$ and $p_w(1/a) \in R_{\bar{v}} \Longrightarrow 1/a \in R$. Furthermore, $a \notin R_w \Longrightarrow 1/a \in M_w \Longrightarrow p_w(1/a) = 0 \Longrightarrow 1/a \in R$. Therefore, by 2.14, $R = R_v$, where v is a valuation of K. Since $R_{\bar{v}} \neq D_w^x$ we have $R_w \supset R_v$ and $R_w \neq R_v$, i.e., $w \longrightarrow v$ and $w \neq v$. Therefore $v = w \circ v^*$, where v^* is the valuation of D_w^x with $R_{v^*} = p_w(R_v) = p_w(R) = R_{\bar{v}}$ so that $v^* = \bar{v}$ and $v = w \circ \bar{v}$.

PROPOSITION 2.34. Let A be a subdomain of a field K and P a prime ideal in A. Let $R = A_P$, $M = PR$, and S = the integral closure of R in K. Let W be the set of valuations of K having center P in A. Then $S \neq K \Longleftrightarrow W$ is nonempty $\Longrightarrow S = \bigcap_{v \in W} R_v$.

PROOF. First observe that W = the set of valuations of K having center M in R. If we include the trivial valuation in W then it is enough to prove $S = \bigcap_{v \in W} R_v$. Let W^* be the set of all valuations of K, including the trivial one, whose valuation rings contain R. Then by 2.25, $S = \bigcap_{v \in W^*} R_v \subset \bigcap_{v \in W} R_v$. Hence it is enough to prove that $w \in W^* \Longrightarrow$ there exists $v \in W$ with $w \longrightarrow v$. Let $N = R \cap M_w$. Then $N \subset M$. If $N = M$, nothing to prove. So assume $N \neq M$. Let $p_w(R) = \bar{R}$ and $p_w(M) = \bar{M}$. Then $\bar{M}$ is a nonzero prime ideal in $\bar{R}$, (in fact, $\bar{M}$ is the unique maximal ideal in $\bar{R}$) and hence by 2.22 there exists a valuation $\bar{v}$ of D_w^x having center $\bar{N}$ in $\bar{R}$. Let $v = w \circ \bar{v}$. Then $v \in W$ and $w \longrightarrow v$.

LEMMA 2.35. Let K be a field and K^* and overfield. Let v and v^* be valuations of K and K^* respectively. Then v^* is an extension of $v \Longleftrightarrow v^*$ has center M_v in R_v and we may consider S_v

to be a subgroup of S_{v*}.

PROOF. ($\Longrightarrow$) Since $v = v* \mid K^x$, $S_v = v(K^x)$ which is a subgroup of S_{v*} since K^x is a subgroup of $(K*)^x$. Also $R_v - (0) = v^{-1}(S_v^+ \cup (0)) = (v*)^{-1}(S_v^+ \cup (0)) \cap K^x = (v*)^{-1}(S_{v*}^+ \cap (0)) \cap K^x = R_{v*} - (0) \cap K$. $M_v - (0) = v^{-1}(S_v^+) = (v*)^{-1}(S_{v*}^+) \cap K^x = M_{v*} - (0) \cap K$.

($\Longleftarrow$) Let v' be the restriction of $v*$ to K. Then v' has center M_v in R_v and hence by 2.31 (2), $v' = v$.

PROPOSITION 2.36. Let K be a field and $K*$ an overfield of K. Let v be a valuation of K, S = the integral closure of R_v in $K*$ and W = the set of extensions of v to $K*$. Then $S = \bigcap_{v* \in W} R_{v*}$.

PROOF. A direct consequence of 2.34 and 2.35.

LEMMA 2.37. Let v be a valuation of a field K, $K*$ an algebraic extension of K, $v*$ an extension of v to $K*$, S = the integral closure of R_v in $K*$, $P = S \cap M_{v*}$ and $R* = S_P$. Then $R* = R_{v*}$.

PROOF. Now $R_{v*} \supset R_v$ and hence $R_{v*} \supset S$. Therefore $R_{v*} \supset R*$. So it remains to be shown that $a \in R_{v*} \Longrightarrow a \in R*$. We may assume $a \neq 0$. $K*/K$ algebraic $\Longrightarrow$

$$k_o a^n + \dots + k_n = 0, \quad \text{with } k_i \in K, \quad \text{with } k_o \neq 0 \ .$$

Let $t = \min (v(k_o), \dots, v(k_n))$. Let j be minimum with $v(k_j) = t$. Let $h_i = k_i/k_j$. Then

$$h_o a^n + \dots + h_{j-1} a^{n-j+1} + a^{n-j} + h_{j+1} a^{n-j-1} + \dots + h_n = 0$$

and $v(h_i) > 0$ if $i < j$ and $v(h_i) \geq 0$ if $i > j$. Dividing out by a^{n-j} we get

$$\left[h_o a^j + h_1 a^{j-1} + \dots + h_{j-1} a + 1\right] +$$

$$+ \frac{1}{a}\left[h_{j+1} + h_{j+1}\left(\frac{1}{a}\right) + \dots + h_n \left(\frac{1}{a}\right)^{n-j-1}\right] = 0 \ .$$

Let b = the sum in the first bracket, and
c = the sum in the second bracket.

Let v' be an arbitrary extension (2.23) of v to $K*$. Then $[v'(a) \geq 0 \Longrightarrow v'(b) \geq 0 \Longrightarrow v'(c) = v'(-ab) \geq 0]$ and $[v'(a) < 0 \Longrightarrow v'(c) \geq 0 \Longrightarrow v'(b) = v'(-c/a) \geq 0]$, i.e., $v'(b) \geq 0$ and $v'(c) \geq 0$. Therefore, by 2.36, $b, c \in S$. Now $v*(a) \geq 0$, and hence $v*(b) = 0$, so that $b \notin P$. Hence $a = -c/b \in S_P = R*$.

11. Ramification theory of valuations.

PROPOSITION 2.38. Let K be a field, v a valuation of K, K^* an algebraic extension of K, (R_1^*, M_1^*), (R_2^*, M_2^*), ... the local rings in K^* lying above R_v. Then R_i^* is the valuation ring of a valuation v_i^* of K^* and $v_1^*, v_2^*, \ldots$ are exactly the extensions of v to K^*.

PROOF. In 2.37, we have proved that the valuation rings of the extensions of v to K^* must be among the rings R_i^*. To prove the converse: By 2.22, there exists a valuation v_i^* of K^* having center M_i^* in R_i^*. Let S be the integral closure of R_v in K^* and let $P_i = M_i^* \cap S$. Then v_i^* has center P_i in S and center M_v in R_v so that, by 2.35, v_i^* is an extension of v and hence by 2.37, $R_i^* = R_{v_i^*}$.

DEFINITION 2.39. Let K^* be a galois extension of a field K and let v be a valuation of K. Let $R_1^*, \ldots, R_t^*$ be the local rings in K^* lying above R_v. Then by 2.38 $R_j^* = R_{v_j^*}$ where $v_1^*, \ldots, v_t^*$ are exactly the extensions of v to K^*. Hence we can apply the entire considerations of Section 7.

We define:

$G^s(v_j^*/v)$ = (the splitting group of v_j^* over v) = $G^s(R_{v_j^*}/R_v)$

$F^s(v_j^*/v)$ = (the splitting field of v_j^* over v) = $F^s(R_{v_j^*}/R_v)$

$G^i(v_j^*/v)$ = (the inertia group of v_j^* over v) = $G^i(R_{v_j^*}/R_v)$

$F^i(v_j^*/v)$ = (the inertia field of v_j^* over v) = $F^i(R_{v_j^*}/R_v)$.

PROPOSITION 2.40. Let K^*/K be a galois extension, v^* a valuation of K^* and v the K-restriction of v^*. Let $K^s = F^s(v^*/v)$ and let v^s be the K^s-restriction of v^*. Then $S_{v^s} = S_v$.

REMARK 2.41. We shall not prove this proposition here but we shall only show how it follows from a result of Krull. In [K3] Theorem 21 Krull has proved that there exists an intermediate field K_z of K and K^* such that if v_z is the K_z-restriction of v^* to K_z, then v^* is the only K^*-extension of v_z and $S_{v_z} = S_v$. It then follows by 1.46 that $K_z \supset K^s$ and hence $S_v \subset S_{v^s} \subset S_{v_z}$; therefore $S_{v^s} = S_v$. The proof given in [K3] is quite involved and hence we pose the following question:

Is it possible to prove that $S_{v^s} = S_v$ from our Theorem 1.47: $M_{v^s} = M_v R_{v^s}$?

PROPOSITION 2.42. Let the notation be as in 2.40. Let $K^i = F^i(v^*/v)$ and let v^i be the K^i-restriction of v^*. Then $S_{v^i} = S_v$.

PROOF. Replacing K by K^s, we may assume that $K = F^s(v^*/v)$. By 1.48, D_{v^i}/D_v is separable and hence has a primitive element $\bar{u}$. Fix u in R_{v^i} in the residue class $\bar{u}$. Then u/R_v is integral and hence by 1.16 the minimal monic polynomial $u^n + a_1u^{n-1} + \dots + a_n = 0$ of u/K has coefficients in R_v. Then $\bar{u}^n + \bar{a}_1\bar{u}^{n-1} + \dots + \bar{a}_n = 0$, where $\bar{a}_i$ is the residue class of a_i mod M_v. Hence $n \geqq [D_{v^i} : D_v]$. By 1.48, $[K^i : K] = [D_{v^i} : D_v]$ and hence u is a primitive element of K^i/K. Given $0 \neq y \in K^i$, we may thus write

$$y = x_0^* + x_1^*u + \dots + x_{n-1}^*u^{n-1}, \quad \text{with} \quad x_t^* \notin K \ .$$

Chooose j such that $v(x_j^*) \leqq v(x_t^*)$ for $t = 0, 1, \dots, n-1$. Let $x_t = x_t^*/x_j^*$. Then with $x_t \in R_v$ for $t = 0, 1, \dots, n$ and $x_j \notin M_v$. Since $\bar{u}$ is of degree n over D_v, we must have

$$x_0 + x_1u + \dots + x_{n-1}u^{n-1} \not\equiv 0 \pmod{M_v}$$

and hence

$$v(x_0 + x_1u + \dots + x_{n-1}u^{n-1}) = 0 \ .$$

Therefore

$$v^*(y) = v^*(x_j^*) = v(x_j^*) \in S_v \ .$$

PROPOSITION 2.43. Let the notation be as in 2.40 and assume that D_v is of zero characteristic. Then $G^i(v^*/v)$ is isomorphic to S_{v^*}/S_v (and hence $G^i(v^*/v)$ is abelian).

REMARK 2.44. This is Satz 3 of Krull [2] (although Krull assumes $\rho(v) = 1$, the proof works in the general case), the proof is quite involved and will be omitted.

COROLLARY 2.45. Let K^*/K be a finite algebraic extension, v a valuation of K and v^* a K^*-extension of K. Then S_{v^*}/S_v is finite.

If D_v is of characteristic zero and K^*/K is separable then the proof follows (by passing to a Galois extension of k^* containing K) from 2.40, 2.42 and 2.43. In the general case, we can prove the stronger assertion; $[S_{v^*} : S_v] \leqq [K^* : K]$ directly as follows:

Let $\bar{u}_1, \dots, \bar{u}_t$ be elements in S_{v^*} lying in distinct S_v-cosets. Fix $u_i \in K^*$ with $u^*(u_i) = \bar{u}_i$. Suppose, if possible, there exist elements a_i not all zero in K such that $a_1u_1 + \dots + a_tu_t = 0$. Then for some $i \neq j$ we must have $a_i \neq 0 \neq a_j$ and $v^*(a_iu_i) = v^*(a_ju_j)$, i.e., $v^*(u_i) = v^*(a_j/a_i)v^*(u_j)$, i.e., $\bar{u}_i = v(a_j/a_i)\bar{u}_j$, i.e., $\bar{u}_i$ and

$\bar{u}_j$ are in the same S_v-coset which is a contradiction. Therefore $u_1, \ldots, u_t$ are linearly independent over K. Therefore $[S_{v*} : S_v] \leqq [K* : K]$.

DEFINITION 2.45A. In the above notation we define

$$r(v*/v) = \text{ramification index of } v* \text{ over } v = [S_{v*} : S_v] .$$

12. Valuations of algebraic function fields. Let K be a field and $K*$ an overfield of K, let v be a valuation of K and $v*$ an extension of v to $K*$ where v and $v*$ are allowed to be trivial. We define

$$(v\text{-dimension of } v*) = (\text{the transcendence degree of } D_{v*}/D_v) .$$

If $K*$ is a finitely generated extension of K of transcendence degree s, then $K*$ is said to be an algebraic function field of dimension s over the ground field K.

PROPOSITION 2.46. Let K be a field and $K*$ an extension of K of finite transcendence degree s. Let $v*$ be a valuation of $K*$ and let v be the K-restriction of $v*$ where we allow v to be trivial. Let d be the v-dimension of $v*$. Let r and $r*$ be the rational ranks of v and $v*$ and let ρ and $\rho*$ be the ranks of v and $v*$ respectively. Then we have the following: (1) If $r*$ is finite, then $r* + d \leqq r + s$. (2) If v is an integral direct sum, if $K*/K$ is finitely generated, and if $r* + d = r + s$, then $v*$ is an integral direct sum and R_{v*}/M_{v*} is finitely generated over R_v/M_v (note that $R_v/M_v = K$ if v is trivial). (1*) If ρ is finite, then $\rho* + d \leqq \rho + s$. (3) If v is discrete, if $K*/K$ is finitely generated, and if $\rho* + d = \rho + s$, then $v*$ is discrete and R_{v*}/M_{v*} is finitely generated over R_v/M_v.

PROOF. First assume that r is finite. We begin by proving the weaker inequality

$$\text{(A)} \qquad r* \leqq r + s .$$

Suppose $s = 0$. Given $0 \neq u \in K*$, let $f(X) = a_0X^n + a_1X^{n-1} + \ldots + a_n$, $a_0 = 1$, be the minimal monic polynomial of u over K. Since $f(u) = 0$, there exist distinct integers i and j such that $v*(a_iu^{n-i}) = v*(a_j^{n-j}) \neq \infty$ and hence $v*(u) = v(a_i/a_j)/(i - j)$, i.e., the value of u depends rationally on the value of $(a_i/a_j) \in K$. Therefore $r* = r = r + s$. Now suppose $s > 0$ and assume that (A) is true for $s - 1$. Let $z_1, z_2, \ldots, z_{s-1}$ be part of a transcendence basis of $K*/K$. Let

$K_1 = K(z_1, z_2, \ldots, z_{s-1})$, let v_1 be the restriction of v^* to K_1 (v_1 may be trivial), and let r_1 be the rational rank of v_1. By our induction hypothesis, $r_1 \leq r + s - 1$. If the value of every nonzero element of K^* is rationally dependent on the values of elements of K_1, then $r^* = r_1 \leq r + s - 1 \leq r - s$, and we are through. Now suppose that there is a nonzero element z in K^* such that $h = v^*(z)$ does not depend rationally on the values of elements of K_1. Then, by the $s = 0$ case, z is transcendental over K_1. Let $f(X) = f_o + f_1X + \ldots + f_nX^n$ and $g(X) = g_o + g_1X + \ldots + g_{n^*}X^{n^*}$ be nonzero elements of $K_1[X]$. Let $a_i = v^*(f_i)$ if $f_i \neq 0$ and $b_i = v^*(g_i)$ if $g_i \neq 0$. Since h depends rationally neither on the a_i nor on the b_i, there exist integers p and q such that $\infty \neq v^*(f_pz^p) < v^*(f_iz^i)$ whenever $i \neq p$ and $f_i \neq 0$, and $\infty \neq v^*(g_qz^q) < v^*(g_iz^i)$ whenever $i \neq q$ and $g_i \neq 0$; i.e., $v^*(f(z)/g(z)) = v^*(f_p/g_q) + (p - q)h$. *Thus, the value of any nonzero element of* $K_1(z)$ *is of the form* $a + mh$ *where* a *is in the value group of* v_1 *and* m *is an integer*, i.e., if r_2 is the rational rank of the restriction of v^* to $K_1(z)$ (this restriction may be trivial), then

$$r_2 = r_1 + 1 \leq r + (s - 1) + 1 = r + s \quad .$$

Since $K^*/K_1(z)$ is an algebraic extension, by the case $s = 0$, we have $r^* = r_2 \leq r + s$. Thus the induction is complete and (A) has been proved. *Also observe that if* v_2 *is the restriction of* v^* *to* $K_2 = K_1(z)$, *then the residue fields of* v_1 *and* v_2 *coincide*. For, in the above notation, since $v_2(f_iz^i) > v_2(f_pz^p)$ whenever $i \neq p$ and $f_i \neq 0$, we must have that $f(z)/(f_pz^p)$ belongs to R_{v_2} and that

$$f(z)/(f_pz^p) = 1 + \sum_{i \neq p} (f_i/f_p)z^{i-p} \equiv 1 \pmod{M_{v_2}} \quad .$$

Similarly, $g(z)/(g_qz^q)$ belongs to R_{v_2} and $g(z)/(g_qz^q) \equiv 1 \pmod{M_{v_2}}$. Now assume that $f(z)/g(z)$ belongs to R_{v_2}. We want to show that we can find e in R_{v_1} with $f(z)/g(z) \equiv e \pmod{M_{v_2}}$. If $f(z)/g(z)$ belongs to M_{v_2}, we can take $e = 0$. Now suppose that $f(z)/g(z)$ does not belong to M_{v_2}, then $0 = v_2(f(z)/g(z)), v_2(f_p/g_q) + (p - q)h$ and since h does not depend rationally on $v_2(f_p/g_q)$, we must have $p - q = 0$, i.e., that $p = q$ and $v_2(f_p/g_p) = 0$. Let $e = f_p/g_p$. Then

$$f(z)/g(z) = (f_p/g_p)(f(z)/f_pz^p)(g_pz^p/g(z)) \equiv e \pmod{M_{v_2}} \quad ,$$

since $f(z)/f_pz^p$ and $g_pz^p/g(z)$ are both congruent to one modulo M_{v_2}.

This proves our second underlined assertion.

To prove (1), let us retain our assumption that r is finite, and let $\bar{y}_1, \bar{y}_2, \ldots, \bar{y}_d$ be a transcendence basis of R_{v*}/M_{v*} over R_v/M_v and fix y_i in R_{v*} belonging to the residue class $\bar{y}_i$. Let $K' = K(y_1, y_2, \ldots, y_d)$ and let v' be the restriction of $v*$ to K'. Given $0 \neq f(X_1, X_2, \ldots, X_d)$ in $K[X_1, X_2, \ldots, X_d]$, choose a coefficient q of f having minimum v-value and let $F(X_1, X_2, \ldots, X_d) = (1/q)f(X_1, X_2, \ldots, X_d)$. Then all the coefficients of $F(X_1, X_2, \ldots, X_d)$ belong to R_v and at least one of them is equal to 1. Let $\bar{F}(X_1, X_2, \ldots, X_d)$ be the polynomial gotten by reducing the coefficients of $F(X_1, X_2, \ldots, X_d)$ modulo M_v. Since $\bar{F}(X_1, X_2, \ldots, X_d)$ has a coefficient equal to 1 and since $\bar{y}_1, \bar{y}_2, \ldots, \bar{y}_d$ are algebraically independent over R_v/M_v, we must have $\bar{F}(\bar{y}_1, \bar{y}_2, \ldots, \bar{y}_d) \neq 0$, i.e., $v*(F(y_1, y_2, \ldots, y_d)) = 0$, i.e., $v*(f(y_1, y_2, \ldots, y_d)) = v(q) \neq \infty$, and hence $f(y_1, y_2, \ldots, y_d) \neq 0$. Thus $y_1, y_2, \ldots, y_d$ are algebraically independent over K and <u>the value groups of v and v' are identical</u>. Since the transcendence degree of $K*/K'$ is $s - d$, (1) follows by applying (A) to $K*/K'$. Now let $g(X_1, X_2, \ldots, X_d)$ and $h(X_1, X_2, \ldots, X_d)$ be arbitrary nonzero elements of $K[X_1, X_2, \ldots, X_d]$ and let

$$y = f(y_1, y_2, \ldots, y_d)/g(y_1 \cdot y_2, \ldots, y_d) \quad .$$

Fix coefficients a and b of g and h respectively having minimum v'-values, and let $p = a/b$. Let $G(X_1, X_2, \ldots, X_d) = (1/a)g(X_1, X_2, \ldots, X_d)$ and $H(X_1, X_2, \ldots, X_d) = (1/b)h(X_1, X_2, \ldots, X_d)$. Then, as above,

$$v'(g(y_1, y_2, \ldots, y_d)/h(y_1, y_2, \ldots, y_d) = v'(a/b) \quad .$$

Hence $y = g(y_1, y_2, \ldots, y_d)/h(y_1, y_2, \ldots, y_d)$ belongs to $R_{v'}$ if and only if $p = a/b$ belongs to $R_{v'} \cap K = R_v$. Now assume that y does belong to $R_{v'}$. Let $\bar{y}$ and $\bar{p}$ be the residue classes modulo $M_{v'}$ containing y and p respectively. Let $\bar{G}(X_1, X_2, \ldots, X_d)$ and $\bar{H}(X_1, X_2, \ldots, X_d)$ be the polynomials obtained respectively from $G(X_1, X_2, \ldots, X_d)$ and $H(X_1, X_2, \ldots, X_d)$ by reducing their coefficients modulo $M_{v'}$. Since $\bar{H}$ has a coefficient equal to one and since $\bar{y}_1, \bar{y}_2, \ldots, \bar{y}_d$ are algebraically independent over R_v/M_v, we have that

$$\bar{y} = \bar{p}\bar{G}(\bar{y}_1, \bar{y}_2, \ldots, \bar{y}_d)/\bar{H}(\bar{y}_1, \bar{y}_2, \ldots, \bar{y}_d) \quad .$$

<u>Therefore</u> $R_{v'}/M_{v'} = (R_v/M_v)(\bar{y}_1, \bar{y}_2, \ldots, \bar{y}_d)$, <u>and hence in particular</u> $R_{v'}/M_{v'}$ <u>is finitely generated over</u> R_v/M_v.

Now assume that v is an integral direct sum, that $K*/K$ is finitely generated, and that $r^* + d = r + s$. Let K' and v' be as above. Then v and v' have the same value groups, K^*/K' is a finitely generated extension of transcendence degree $e = s - d = r^* - r$, and $R_{v'}/M_{v'}$ is finitely generated over R_v/M_v. Fix an integral basis $t_1, t_2, \ldots, t_r$ of the value group of v'. Let $x_1, x_2, \ldots, x_e$ be a transcendence basis of K^*/K'. Let $K_i' = K'(x_1, x_2, \ldots, x_i)$, v_i' = the restriction of v^* to K_i', and r_i' = the rational rank of v_i'. Since $r^* = r + e$, we must have, in view of (1), $r_i = r_{i-1} + 1$ for $i = 1, 2, \ldots, e$. Let $v^*(x_i) = t_{r+i}$. By applying the first of the above underlined remarks successively to the extensions K_1'/K', K_2'/K_1', $\ldots$, K_e'/K_{e-1}', we conclude that for any nonzero element x of K_e' we have

$$v^*(x) = a + m_{r+1}t_{r+1} + m_{r+2}t_{r+2} + \cdots + m_{r^*}t_{r^*} ,$$

where a is the value of an element of K' and where $m_{r+1}, m_{r+2}, \ldots, m_{r^*}$ are integers; since $a = m_1t_1 + m_2t_2 + \cdots + m_rt_r$ where $m_1, m_2, \ldots, m_r$ are integers, we finally have: $v^*(x) = m_1t_1 + m_2t_2 + \cdots + m_{r^*}t_{r^*}$. Therefore v_e' is an integral direct sum. Since K^*/K_e' is a finite algebraic extension, the value group of v_e' is a subgroup of the value group of v^* of finite index and hence v^* is an integral direct sum. Now by the second underlined remark above, the residue field of v_e' coincides with the residue field of v'. Since the residue field of v' is finitely generated over the residue field of v and since K^*/K_e' is a finite algebraic extension, we conclude that R_{v^*}/M_{v^*} is finitely generated over R_v/M_v. This proves (2).

The proof of (1*) is entirely similar to that of (1). Finally, assume that v is discrete, $\rho^* + d = \rho + s$, and that K^*/K is finitely generated. The discreteness of v implies that $\rho = r$. Since by (1), $r^* + d \leqq r + s$ and since $r^* \geqq \rho^*$, it follows that $r^* = \rho^*$ and that $r^* + d = r + s$. Hence by (2), v^* is an integral direct sum and R_{v^*}/M_{v^*} is finitely generated over R_v/M_v. Since $r^* = \rho^*$, v^* is discrete. This proves (3).

COROLLARY 2.47. In the notation of the above lemma, assume that v is trivial, i.e., that v^* is a valuation of K^*/K. Then: (1) $\rho^* + d \leqq r^* + d \leqq s$. Furthermore, if K^*/K is finitely generated (i.e., if K^*/K is an algebraic function field of dimension s), then we have the following: (2) If $r^* + d = s$, then v^* is an integral direct sum and R_{v^*}/M_{v^*} is finitely generated over K. (3) If $\rho^* + d = s$, then v^* is discrete and R_{r^*}/M_{r^*} is finitely generated over K. (4) If $d = s - 1$, then v^* is real discrete and R_{v^*}/M_{v^*} is finitely generated over K.

DEFINITION 2.48. In the notation of 2.47, if K^*/K is finitely generated and if $d = s - 1$, then we shall say that v is a prime divisor of K^*/K.

COROLLARY 2.49. Let K^*/K be a two-dimensional algebraic function field and let v be a valuation of K^*/K. By 2.47, it follows that v is of one of the following four types: (1) v is a prime divisor of K^*/K, i.e., v is real discrete and D_v/K is a one-dimensional algebraic function field; (2) v is discrete of rank two; (3) v is real of rational rank one, but v is not a prime divisor of K^*/K; (4) v is real of rational rank two; in this case v is necessarily an integral direct sum. In cases (2) and (4), D_v/K is finite algebraic, and in case (3), D_v/K is algebraic.

CHAPTER III: NOETHERIAN LOCAL RINGS

For proofs of some of the propositions of Section 13 we shall refer to Chapters III and IV of Northcott's Ideal Theory, [N].

13. The dimension of noetherian local rings. Let P be a prime ideal in a ring A. The upper bound of integers n for which there exist prime ideals $P_1, \ldots, P_n$ in A with $P > P_1 > P_2 \cdots > P_n$ is called the A-*rank* of P, in symbols: $\operatorname{rank}_A P$. The upper bound of integers n for which there exist prime ideals $P_1, \ldots, P_n$ in A with $P < P_1 < \cdots < P_n$ is called the A-*dimension* of P, in symbols: $\dim_A P$. When the reference to A is clear, $\operatorname{rank}_A P$ and $\dim_A P$ will be replaced by $\operatorname{rank} P$ and $\dim P$ respectively. Observe that $\operatorname{rank}_A P = \operatorname{rank}_{A_P} PA_P$ and $\dim_A P = \dim_{A/P}(0)$.

PROPOSITION 3.1. In a noetherian ring A the rank of any prime ideal is finite.

PROOF. Theorem 7, page 60 of [N].

PROPOSITION 3.2. Let A be a domain which is of finite transcendence degree n over a subfield k. Then (1) for any prime ideal P of A we have $\operatorname{rank} P + \dim P \leqq n$. (2) If A is finitely generated over k, then the equality holds.

PROOF. (1) Since $\operatorname{rank}(0) \geqq \operatorname{rank} P + \dim P$, it is enough to prove that $\operatorname{rank}(0) \leqq n$. Let $0 = P_0 < P_1 < \cdots < P_t$ be a chain of prime ideals in A. We have to show that $t \leqq n$. Now if f is a k-homomorphism of A onto a domain B then $\text{tr. deg. } B/k \leqq \text{tr. deg. } A/k$ and the equality holds if and only if f is an isomorphism. [PROOF. Let $u_1, \ldots, u_t$ be algebraically independent elements of B/k and fix v_i in A with $f(v_i) = u_i$, then $v_1, \ldots, v_t$ are algebraically independent over k, hence the inequality. Now assume that the equality holds: let $u_1, \ldots, u_n$ be a transcendence basis of B/k and fix v_i in A with $f(v_i) = u_i$; then $v_1, \ldots, v_n$ is a transcendence basis of A/k. Suppose if possible that for some $0 \neq a \in A$, $f(a) = 0$. Multiplying the minimal monic polynomial of $a/k(v_1, \ldots, v_n)$ by the product of the denominators of the nonzero coefficients, we obtain: $p_s a^s + p_{s-1}a^{s-1} + \cdots + p_o = 0$ with $p_i \in k[v_1, \ldots, v_n]$ and $p_o \neq 0$. Applying f we get: $f(p_o) = 0$, which contradicts the k-algebraic independence of $u_1, \ldots, u_n$.] Therefore, for any non-zero prime ideal Q in A we have: $[\text{tr. deg } (A/Q)/k] \leq n - 1$. Therefore: $[\text{tr. deg } (A/P_i)/k] \leq [\text{tr. deg } (A/P_{i-1})/k] - 1$, for $i = 1, \ldots, t$. Hence $0 \leq [\text{tr. deg } (A/P_t)/k] \leq [\text{tr. deg } (A/P_o)/k] - t = n - t$, i.e., $t \leq n$.

(2) Observe that we can choose a transcendence basis $x_1, \ldots, x_n$ of A/k such that A is integral over $B = k[x_1, \ldots, x_n]$, (this is Noether's normalization theorem, for a proof see [Z1], section 2). Secondly, observe that it follows from 1.22, 1.24 and 1.24B that $\text{rank}_A P = \text{rank}_B P \cap B$ and $\dim_A P = \dim_B P \cap B$. Now we shall make induction on rank P. If rank $P = 0$, then $P = P \cap B = 0$; and since $Q_i = (x_1, \ldots, x_i)B$ is a prime ideal in B with $0 < Q_1 < \cdots < Q_n$ we have $\dim_A(0) = \dim_B(0) \geq n$ and hence, by (1), $\text{rank}_A(0) + \dim_A(0) = \dim_A(0) = n$. Now consider the case of rank $P = 1$. Fix $0 \neq f(x_1, \ldots, x_n)$ in $Q = P \cap B$. Since Q is prime, it contains an irreducible factor $g(x_1, \ldots, x_n)$ of f. Since gB is prime and since $\text{rank}_B Q = 1$, we must have $Q = gB$. By relabeling the x_i we may assume that x_n actually occurs in g. Let $\bar{x}_i$ be the Q-residue class of x_i. Then $\bar{x}_1, \ldots, \bar{x}_{n-1}$ are algebraically independent over k and $\bar{x}_n$ is algebraically dependent over $k[\bar{x}_1, \ldots, \bar{x}_n]$. Hence, $\text{tr. deg } (B/Q)/k = n - 1$. Since A is integral over B and and since $Q = P \cap B$, A/P is an algebraic extension of B/Q and hence $\text{tr. deg } (A/P)/k = n - 1$. Therefore $\dim_A P = \dim_{A/P}(0) = n - 1$ so that $\text{rank}_A P + \dim_A P = n$.

Now assume that rank $P = r > 1$ and assume that (2) is true for all smaller values of rank P. Let $P_1 < \cdots < P_r < P$ be a chain of prime ideals in A. Then rank $P_r = r - 1$ and hence by the induction hypothesis $\text{tr. deg } (A/P_r)/k = \dim_{A/P_r}(0) = \dim_A P_r = n - (r - 1) = n - r + 1$. Since $\text{rank}_{A/P_r} P/P_r = 1$ we have:

$$\begin{aligned}\dim_A P = \dim_{A/P_r} P/P_r &= [\text{tr. deg } (A/P_r)/k] - 1 \\ &= (n - r + 1) - 1 = n - r.\end{aligned}$$

DEFINITION 3.3. Let (R, M) be a noetherian local ring. Then M is generated by a finite number of elements. We define

$\dim R$ = dimension of R
= minimum value of n such that there exist nonzero elements $a_1, \ldots, a_n$ in R for which $(0, a_1, \ldots, a_n)R$ is primary for M.

Let $\dim R = d$. If $a_1, \ldots, a_d$ are nonzero elements in R such that $(0, a_1, \ldots, a_d)R$ is primary for M, then $(a_1, \ldots, a_d)$ are said to form a system of <u>parameters</u> in R. Observe that M/M^2 can be considered as a vector space over the field R/M. We define:

$$\begin{aligned} \operatorname{emdim} R &= \text{embedding dimension of } R \\ &= \dim_{R/M} M/M^2 \\ &= \text{the } (R/M)\text{-dimension of the vector space } M/M^2. \end{aligned}$$

Observe that $\operatorname{emdim} R \geq \dim R$.

LEMMA 3.4. Let (R, M) be a noetherian local ring and let H be an ideal in R. Then $\bigcap_{i=1}^{\infty}(H + M^i) = H$. In particular $\bigcap_{i=1}^{\infty} M^i = (0)$.

PROOF. [N], Corollary 1, on page 65.

PROPOSITION 3.5. Let (R, M) be a noetherian local ring. Then (1) $\dim R = \operatorname{rank}_R M$; (2) Every irredundant basis of M contains exactly $(\operatorname{emdim} R)$ elements.

PROOF. [N], Theorem 1 on page 63, and Proposition 6 on page 69.

REMARK 3.6. Let (R, M) be a noetherian local ring of dimension d. Then $d = 0 \iff M$ is the only prime ideal of $R \iff (0)$ is primary for $M \iff M$ consists entirely of nilpotents; also observe that R is a field $\iff M = (0)$ (and hence $d = 0$).

From now on, by a noetherian local ring, unless otherwise stated, we shall mean a noetherian local ring of positive dimension.

DEFINITION 3.7. Let R be a local ring. R is said to be <u>regular</u> if R is noetherian and $\dim R = \operatorname{emdim} R$. [We remark that Auslander-Buchsbaum-Serre have proved that the Homdim (Homological dimension of) R is finite $\iff R$ is regular $\implies \operatorname{Homdim} R = \dim R$.]

PROPOSITION 3.8. Let (R, M) be an n-dimensional noetherian local ring, and let $x_1, \ldots, x_m$ be a minimal basis of M. Then R is regular if and only if one of the following is true:

(1) If $f(X_1, \ldots, X_m)$ is a form of degree s with coefficients in R not all in M then $f(x_1, \ldots, x_m) \notin M^{s+1}$.

(2) If $f(X_1, \ldots, X_m)$ is a form (of arbitrary degree) with coefficients in R not all in M then $f(x_1, \ldots, x_m) \neq 0$.

Now assume that R is regular so that $n = m$. Then R is a normal domain. Let $P_i = (x_1, \ldots, x_i)R$. Then for $i = 0, 1, \ldots, n - 1$, P is a prime ideal in R, rank $P_i = i$, $\dim P_i = n - i$, R/P_i is a regular local ring of dimension $n - i$, and R_{P_i} is a regular local ring of dimension i.

PROOF. ([N], Section 4.6) + (Exercise).

DEFINITION 3.9. Let (R, M) be a regular local ring and K the quotient field of R. For $0 \neq z \in R$, let $\delta(z)$ be the nonnegative integer such that

$$z \in M^{\delta(z)} \quad \text{and} \quad z \notin M^{\delta(z)+1} \quad (\delta(z) \text{ exists by } 3.4).$$

We call $\delta(z)$ the <u>leading degree</u> of z. We set $\delta(0) = \infty$. For z_1, z_2 in R we have $\delta(z_1 z_2) = \delta(z_1) + \delta(z_2)$ [for proof, see Lemma 3 on page 70 of N], hence if we set $\delta(z_1/z_2) = \delta(z_1) - \delta(z_2)$, δ becomes a real discrete valuation of K, and we call this the <u>M-adic valuation</u> of K.

Now let T be a representative system of R/M in R and fix a minimal basis $x_1, \ldots, x_n$ of M. Then given $0 \neq z \in R$, we can write $z = f(x_1, \ldots, x_n)$, where f is a form of degree $\delta(z)$ with coefficients in R not all in M. Let $F(X_1, \ldots, X_n)$ be the form of degree $\delta(z)$ in $T[X_1, \ldots, X_n]$ gotten from f after replacing the coefficients by their T-representatives. Then

$$z - F(x_1, \ldots, x_n) \in M^{\delta(z)+1} \quad .$$

It follows from 3.8 (2) that $F(X_1, \ldots, X_n)$ is the only form of degree $\delta(z)$ in $T[X_1, \ldots, X_n]$ which satisfies the above inclusion. We define

$\tau(z) = F(X_1, \ldots, X_n)$
= the leading form of z with respect to the representative system T of R/M and the minimal basis $x_1, \ldots, x_n$ of M.

DEFINITION 3.10. Let (R, M) be a local domain. If R contains a subfield k and a finite number of elements $a_1, \ldots, a_n$ such that $R = A_P$ where $A = k[a_1, \ldots, a_n]$ and $P = A \cap M$, then we shall say R is <u>algebraic with ground field</u> k. We set: k-rank of R = tr. deg. $(R/M)/k$. Observe that by 3.2: k-rank R + dim R = tr. deg. R/k. R will be said to be <u>algebraic</u> if R is algebraic with some ground field. Observe that an algebraic local ring is noetherian.

LEMMA 3.11. Let R be a normal local domain with quotient field K and let R^* be a local ring lying above R in a finite algebraic extension K^* of K. Then

(1) If K^*/K is separable and R is noetherian, then R^* is noetherian with dim R = dim R^*, and
(2) If R is algebraic with ground field k, then R^* is algebraic with ground field k with k-rank R = k-rank R^* and dim R = dim R^*.

PROOF. If K^*/K is separable everything follows from the fact that the integral closure in K^* of a normal noetherian domain A quotient K is a finite A-module [W2, page 79]. If K^*/K is not separable, then in the proof of (2), one has to invoke the following: If A is a domain with quotient field K such that A is finitely generated over a subfield k, then the integral closure of A in K^* is finitely generated over k [for proof see the last paragraph on page 507 of Z1].

PROPOSITION 3.12. Let (R^*, M^*) be an algebraic local domain with ground field k and quotient field K^* such that $k\text{-rank } R^* = 0$. Let K be a subfield of K^* such that $k \subset K$ and K^*/K is finite algebraic. Then each of the following two is a necessary and sufficient condition for the existence of a local ring R in K lying below R^*:

(1) The ideal $(K \cap M^*)R^*$ is primary for M^*.

(2) There exist a finite number of elements $u_1, \ldots, u_h$ in $K \cap R^*$ such that the ideal $(u_1, \ldots, u_h)R^*$ is primary for M^*.

Furthermore, when R exists, it is algebraic with ground field k.

PROOF. The proof of this is rather involved, and hence we refer to Proposition 1 of [A6].

DEFINITION 3.13. Let (R, M) and (S, N) be local domains with a common quotient field K. If S has center M in R and if there exist a finite number of elements $u_1, \ldots, u_h$ in S such that $S = A_P$ where $A = R[u_1, \ldots, u_h]$ and $P = A \cap N$, then we shall say that S is a <u>finite transform</u> of R.

LEMMA 3.14. Let S be a finite transform of a local domain R.

(1) If R is noetherian, then S is noetherian.

(2) If R is algebraic with ground field k, then S is algebraic with ground field k.

(3) If T is a finite transform of S, then T is a finite transform of R.

PROOF. Exercise.

PROPOSITION 3.15. Let R be a one-dimensional local domain. Then R is regular if and only if it is normal.

PROOF. See Theorem 8 on page 76 of [N].

PROPOSITION 3.16. Let R be a domain. Then

(1) R is the valuation ring of a real discrete valuation $v \iff R$ is a regular one-dimensional local domain.

(2) If R is the valuation ring of a valuation v, then R is noetherian $\iff v$ is real discrete.

(3) If R is a local domain, then R is the valuation ring of a valuation $\Longleftrightarrow$ R is regular of dimension one.

PROOF. (1) ($\Longrightarrow$) Fix x in R with $v(x) = 1$. For an ideal H in R, let $v(H) = \min v(h)$ for h in H. Then $H = x^{v(H)}R$ and in particular $M_v = xR$. Hence R is noetherian (in fact, a principal ideal domain) and M_v is the only prime ideal in R. Hence $\dim R = 1 = \operatorname{emdim} R$. ($\Longleftarrow$) Let M be the maximal ideal of R and let v be the M-adic valuation. Then $R = R_v$.

(2) ($\Longrightarrow$) Let P be a prime ideal in R other than M. Since P is prime, $M^i \subset P$ would imply $M = P$ which is not the case, hence $M^i \not\subset P$. Since the ideals in R_v are simply ordered by inclusion, we must have $P \subset M^i$ for all i. By 3.4, $P \subset \bigcap_{i=1}^{\infty} M^i = (0)$, i.e., $P = (0)$. Therefore v is real. Now v is nondiscrete implies there exist real numbers $a_i \in R_v$ with $a_1 > a_2 > \cdots > 0$, i.e., $H_1 < H_2 < \cdots$, where $H_i = \{r \in R_v \mid v(r) = a_i\}$ is a strictly ascending infinite sequence of ideals in R which contradicts the noetherian character of R. ($\Longleftarrow$) This has been proved in (1).

(3) Follows from (1) and (2).

DEFINITION 3.17. Recall that a domain R is said to be a <u>dedekind domain</u> if

(1) R is normal

(2) R is noetherian, and

(3) Every prime ideal in R is maximal.

It follows that $(R, M_1, \ldots, M_n)$ is a normal noetherian semilocal domain such that $\dim R_{M_i} = 1$ for $i = 1, \ldots, n$ $\Longleftrightarrow$ R is a dedekind domain with a finite number of prime ideals $M_1, \ldots, M_n$.

PROPOSITION 3.18. Let R be a one-dimensional noetherian local domain with quotient field K and let S be the integral closure of R in K. Then S is a principal ideal ring with only a finite number of prime ideals $P_1, \ldots, P_n$ and hence S is the intersection of the real discrete valuation rings $R_{P_1}, \ldots, R_{P_n}$.

PROOF. See [K5] and Section 39 of [K4].

LEMMA 3.18A. Let (R, M) be a two-dimensional algebraic regular local domain with quotient field K, let P be a minimal prime ideal in R, let $\bar{R} = R/P$ and $\bar{M} = M/P$. Let $\bar{K}$ be the quotient field of $\bar{R}$ and let T be the integral closure of $\bar{R}$ in $\bar{K}$. Then T is a finite $\bar{R}$-module.

PROOF. See page 511 of [Z1].

PROPOSITION 3.18B. Let (R, M) be a normal algebraic local

domain with quotient field K, let K^* be a finite separable algebraic extension of K and let (R^*, M^*) be a local ring in K^* lying above R. Assume that R^* is regular, $R^*/M^* = R/M$ and $MR^* = M^*$. Then R is regular.

PROOF. The proof of this proposition follows from the considerations made in Section 2 of [A2]; since these considerations use the notion of the completion of a local ring, we shall omit the proof.

14. Quadratic transforms.

LEMMA 3.19. Let (R, M) be a regular local domain of dimension $s > 1$ with quotient field K and residue field k. Let $\{x_1, x_2, \ldots, x_s\}$ be a system of parameters in R. Let $y_1 = x_1$, $y_i = x_i/x_1$ for $i = 2, 3, \ldots s$; and let $S = R[y_2, y_3, \ldots, y_s]$. Then $y_1S \cap R = M$, and $k = R/M$ can be canonically identified with a subfield of $S/(y_1S)$. Furthermore, the residues $\bar{y}_2, \bar{y}_3, \ldots, \bar{y}_s$ modulo y_1S of $y_2, y_3, \ldots, y_s$ are algebraically independent over k, and $S/(y_1S)$ can be canonically identified with the polynomial ring $k[\bar{y}_2, \bar{y}_3, \ldots, \bar{y}_s]$ in $s - 1$ independent variables.

PROOF. Let v be the M-adic valuation of K. Now $R_v \supset S$ and $v(y_1) > 0$, and hence $1 \notin y_1S$. Therefore, $y_1S \cap R \neq R$. Since $y_1S \supset M$ and since M is the maximal ideal in R, it follows that $y_1S \cap R = M$; and hence we can canonically identify $k = R/M$ with a subfield of $S/(y_1S)$, in the usual fashion. With this identification, it is clear that $S/(y_1S)$ is generated, as a ring over k, by $\bar{y}_2, \bar{y}_3, \ldots, \bar{y}_s$. Note that the center of v in R is the maximal ideal M, while the center of v in S is the minimal prime ideal y_1S. The valuation ring R_v is the quotient ring of S with respect to y_1S.

The only thing that remains to be shown is that $\bar{y}_2, \bar{y}_3, \ldots, \bar{y}_s$ are algebraically independent over k. Assume the contrary. Then there exists a polynomial

$$f(y_2, y_3, \ldots, y_s) = \Sigma f_{i_2,i_3,\ldots,i_s} y_2^{i_2} y_3^{i_3} \ldots y_s^{i_s} ,$$

with the coefficients $f_{i_2,i_3,\ldots,i_s}$ in R, but not all in M, such that $f(y_2, y_3, \ldots, y_s) \equiv 0 \pmod{y_1S}$, i.e.,

$$f(y_2, y_3, \ldots, y_s) = y_1 g(y_2, y_3, \ldots, y_s) ,$$

where $g(y_2, y_3, \ldots, y_s)$ is a polynomial in $y_2, y_3, \ldots, y_s$ with coefficients in R. Multiplying both sides of the above equation by a suitable high power x_1^t of x_1 and setting

$$F_t(x_1, x_2, \ldots, x_s) = \Sigma f_{i_2,i_3,\ldots,i_s} x_1^{t-i_2-i_3-\cdots-i_s} x_2^{i_2} x_3^{i_3} \cdots x_s^{i_s} ,$$

we get

$$F_t(x_1, x_2, \ldots, x_s) = x_1^{t+1} g(y_2, y_3, \ldots, y_s) .$$

Hence $v(F_t(x_1, x_2, \ldots, x_s)) \geq t + 1$, i.e., $F_t(x_1, x_2, \ldots, x_s)$ is contained in M^{t+1}. This is a contradiction, since $F_t(x_1, x_2, \ldots, x_s)$ is a form of degree t in $x_1, x_2, \ldots, x_s$ with coefficients in R, not all in M. Hence $\bar{y}_2, \bar{y}_3, \ldots, \bar{y}_s$ are algebraically independent over k, and this completes the proof.

LEMMA 3.20. Let (R, M) be a regular local domain of dimension $s > 1$ and let $x_1, x_2, \ldots, x_s$ be a minimal basis of M. Let v be a valuation of the quotient field K of R having center M in R. Suppose we have arranged the x_i so that $v(x_1) \leqq v(x_i)$ for $i = 1, 2, \ldots, s$. Let $A = R[x_2/x_1, x_3/x_1, \ldots, x_s/x_1]$, $P = A \cap M_v$, $S = A_P$ and $N = PS$. Then (S, N) is a regular local domain of dimension $t \leqq s$, v has center N in S and if by d and d^* we denote respectively the R-dimension and the S-dimension of v then we have: $s - t = d - d^*$.

PROOF. By 3.19, (A/x_1A) can be canonically identified with a polynomial ring $A^* = (R/M)(X_2, X_3, \ldots, X_s)$ in $s - 1$ variables over R/M. Let $P^* = P/(x_1A)$. Then $A^*_{P^*}$ is a regular local ring [Z2, Section 4.1] of dimension $h \leqq s - 1$. Fix $y_2^*, y_3^*, \ldots, y_{h+1}^*$ in A^* such that

$$P^*A^*_{P^*} = (y_2^*, y_3^*, \ldots, y_{h+1}^*)A^*_{P^*} .$$

Fix y_i in A belonging to the residue class y_i^*. Then it is clear that $(x_1, y_2, y_3, \ldots, y_{h+1})S = N$ and that

$$t = \dim S = \text{rank } N = 1 + \text{rank } P^* = 1 + h .$$

Therefore S is regular. Also $t \leqq 1 + h \leqq (s - 1) + 1 = s$. Since $N \cap R = M$, we may canonically assume that $R/M \subset S/N \subset R_v/M_v$. Since the transcendence degree of $S/N = A^*/P^*$ over R/M is $s - 1 - h = s - t$, we conclude that: $s - t = (R\text{-dimension of } v) - (S\text{-dimension of } v) = d - d^*$.

DEFINITION 3.21. In the notation of the above lemma, S is called "the first (or immediate) quadratic transform of R along v." Let now $R_0 = R$ and let R_i be the first quadratic transform of R_{i-1} along v assuming that $\dim R_{i-1} > 1$. If $\dim R_i > 1$, for

$i = 1, 2, \ldots, n - 1$, then R_n will be defined and we shall say that "R_n is the n-th quadratic transform of R along v." If S is the n-th quadratic transform of R along v for some n, we shall say that "S is a quadratic transform of R along v." Finally, if S is a quadratic transform of R along some valuation v having center M in R, we shall say that "S is a quadratic transform of R."

LEMMA 3.22. Let S be a quadratic transform of a regular local domain R. If R is algebraic with ground field k, then so is S.

PROOF. Follows from 3.14.

CHAPTER IV: TWO-DIMENSIONAL LOCAL DOMAINS

In the next two sections we shall mostly deal with two-dimensional regular local domains which are algebraic. However, all (respectively, almost all) the results of Sections 15 and 16 have been extended by us in [A4] to absolute (respectively, arbitrary) two-dimensional regular local domains.

15. Limits of quadratic sequences.

PROPOSITION 4.1. Let (R_1, M_1), (R_2, M_2), ..., be a sequence of normal local domains with a common quotient field K such that R_{i+1} has center M_i in R_i for $i = 1, 2, \ldots$. Assume that $\cup_{i=1}^{\infty} R_i$ is not the valuation ring of a valuation of K. Then there exist infinitely many valuations w of K which have center M_i in R_i and for which D_w is of positive transcendence degree over R_i/M_i for each i.

PROOF. Let $R = \cup_{i=1}^{\infty} R_i$ and $M = \cup_{i=1}^{\infty} M_i$. We may canonically assume that $R_1/M_1 \subset R_2/M_2 \subset \ldots$. Let $D = \cup_{i=1}^{\infty} R_i/M_i$. Then D is a field and $R/M = D$, i.e., M is a maximal ideal in the domain R. In fact, by 1.5, (R, M) is a local ring. Also observe that R is normal. Since R is not a valuation ring, by 2.14 there exists $x \in K$ with $x \notin R$ and $1/x \notin R$. Let h be the canonical homomorphism of R onto D and X be an indeterminate. For $f(x) = \Sigma_{i=0}^{n} f_i x^i$ with $f_i \in R$ we set $H(f(x)) = \Sigma_{i=0}^{n} h(f_i)X^i$. We assert that H is a homomorphism of $R[x]$ onto $D[X]$.

[PROOF. It suffices to prove that H is single-valued, i.e., that $f(x) = g(x) = \Sigma_{i=0}^{n} g_i x^i \Longrightarrow \Sigma h(f_i)X^i = \Sigma h(g_i)X^i$. Let $q_i = f_i - g_i$. Then we have to show that $q_i \in R$ and $\Sigma_{i=0}^{n} q_i x^i = 0 \Longrightarrow \Sigma_{i=0}^{n} h(q_i)X^i = 0$, i.e., $h(q_i) = 0$ for all i. Assume the contrary. Then for some i, $h(q_i) \neq 0$, i.e., q_i is a unit in R; let t be the maximum value of i for which this is so. Then we can divide the equation $\Sigma q_i x^i = 0$ by q_t and obtain:

$$p_n x^n + p_{n-1} x^{n-1} + \ldots + p_{t+1} x^{t+1} + x^t + p_{t-1} x^{t-1} + \ldots + p_0 = 0$$

with

$$p_i = q_i/q_t \in R, \quad \text{and} \quad h(p_{t+1}) = h(p_{t+2}) = \dots = h(p_n) = 0.$$

Let

$$r = p_n x^{n-t} + p_{n-1} x^{n-t-1} + \dots + p_{t+1} x + 1 \quad ,$$

and

$$s = p_{t-1}(1/x) + p_{t-2}(1/x)^2 + \dots + p_o(1/x)^t \quad ,$$

so that

$$rx^t + x^t s = 0, \quad \text{i.e.,} \quad r = -s \quad .$$

Let v be a valuation of K having center M in R, then $v(x) \geqq 0 \Longrightarrow v(r) \geqq 0$ and $v(x) < 0 \Longrightarrow v(r) = v(-s) > 0$; hence in either case $v(r) \geqq 0$, i.e., $r \in R_v$. Hence by 2.34 $r \in R$. Again by 2.34, $x \notin R \Longrightarrow$ there exists a valuation v of K having center M in R such that $v(x) < 0$. Then $v(1/x) > 0$ so that $v(r) = v(-s) > 0$ and hence $r \in M_v \cap R = M$. Thus

$$(1 - r) + p_{t+1}x + \dots + p_n x^{n-t} = 0 \quad ,$$

with $p_{t+1}, \dots, p_n$ in R and $r \in M$. Now $r \in M \Longrightarrow 1 - r$ is a unit in R and hence dividing the above equation by $(1 - r)x^{n-t}$ we would obtain an equation of integral dependence of $1/x$ over R. This is a contradiction since $1/x \notin R$ and R is normal.]

It is clear that H is the unique extension of h with $H(x) = X$. Let p be any one of the infinitely many prime ideals in $D[X]$. Let $P = H^{-1}(p)$. By 2.22 there exists a valuation w of K having center P in $R[x]$. Since $D_w \supset D(x)$, it is clear that w has the required properties. The infinitely many choices of p give us infinitely many w of the required type.

LEMMA 4.2. Let (R, M) be a two-dimensional local domain with quotient field K. Let P be a minimal prime ideal in R. Then:

(1) R_P is the valuation ring of a real discrete valuation w of K;

(2) There exists at least one and at most a finite number of valuations v of K having center M in R which are composed with w, i.e., for which $R_v \subset R_w$;

(3) If R is algebraic, then each such valuation v is discrete of rank two and R_v/M_v is a finite algebraic extension of R/M (hence in particular v is of R-dimension zero).

PROOF. (1) follows from 1.17 and 3.15. The proof of (2) is as follows:

$R/(R \cap M_w)$ is a local domain of dimension one with quotient field R_w/M_w. Hence, by 3.18, the integral closure of $R/(R \cap M_w)$ in R_w/M_w is a Dedekind domain D with a finite number of prime ideals $P_1, P_2, \ldots, P_h$. Let v_i^* be the real discrete valuation of R_w/M_w with $R_{v_i^*} = D_{P_i}$. Let v_i be the valuation of R_w/M_w which is composed of w and v_i^*. Then $v_1, v_2, \ldots, v_h$ are exactly the valuations described in (2). Finally, (3) follows at once from 2.49.

LEMMA 4.3. Let (R, M) be a two-dimensional algebraic regular local domain with quotient field K. Then:

(1) R is a unique factorization domain, i.e., equivalently, every minimal prime ideal in R is principal; and

(2) There exist infinitely many valuations of K having center M in R which have R-dimension zero and which are discrete of rank two.

PROOF. For the proof of (1), we refer to Theorem 5 of [Z2]. To prove (2) it is enough to show, in view of 4.2, that there exist infinitely many relatively prime irreducible nonunits in R. To show this, let x, y be a minimal basis of M. Let $P = xR$, $\bar{R} = R/P$, $\bar{M} = M/P$, $\bar{K} = R_P/(PR_P)$, $\bar{y}$ = the residue class modulo P containing y, and w = the real discrete valuation of K with $R_w = R_P$. Then $\bar{M} = \bar{y}\bar{R}$ and hence $\bar{R}$ is a regular one-dimensional local domain, i.e., $\bar{R} = R_{\bar{v}}$ where v is a real discrete valuation of $\bar{K}$ with $\bar{v}(\bar{y}) = 1$. Let v be the valuation of K which is composed of w and $\bar{v}$, and let us write the elements of the value group of v as lexicographically ordered pairs of integers. Let $x_m = x + y^m$ where m is a positive integer. Since (x_m, y) is a basis of M, x_m is an irreducible nonunit. Since $v(x_m) = (0, \bar{v}(\bar{y}^m)) = (0, m)$, we have that $v(x_m) \neq v(x_n)$ whenever $m \neq n$. Therefore $x_1, x_2, x_3, \ldots,$ are infinitely many pairwise relatively prime irreducible nonunits in R.

PROPOSITION 4.4. Let (R, M) be an n-dimensional algebraic regular local domain with quotient field K with $n > 1$, and let v be a valuation of K which has center M in R such that Δ_v is of transcendence degree $n - 1$ over R/M. Then the quadratic sequence along v starting from R is necessarily finite, i.e., if $R = R_0$ and if R_i is the first quadratic transform of R_{i-1} along v provided $\dim R_{i-1} > 1$

then for some integer h we have that R_h is one-dimensional; we also have: $R_h = R_v$. Furthermore, there exists a field T with $R/M \subset T \subset R_v/M_v$ such that T is finitely generated over R/M and R_v/M_v is a pure transcendental extension of T of finite positive transcendence degree (we may express this by saying that "R_v/M_v is a ruled extension of R/M").

PROOF. Assume the contrary, i.e., that the quadratic sequence $R = R_o < R_1 < R_2 < \cdots$ along v is infinite. It then follows, by 3.20, that there exists an integer s such that $\dim R_t = \dim R_s$ and R_t-dimension of $v = m - 1$ whenever $t \geq s$, where we have set $m = \dim R_s$. Let $S = \bigcup_{i=o}^{\infty} R_i$ and $N = \bigcup_{i=o}^{\infty} M_i$ where M_i is the maximal ideal in R_i. Then, as in the proof of 4.1, N is the unique maximal ideal in S and $\bigcup_{i=s}^{\infty} R_i/M_i = S/N$. Since, as in the proof of 3.20, R_{t+1}/M_{t+1} is an algebraic extension of R_t/M_t whenever $t \geq s$, it follows that S/N is an algebraic extension of R_t/M_t for any $t \geq s$. Now v has center M_i in R_i for each i and by 2.47, v is real discrete. Suppose, if possible, that S is not the valuation ring of any valuation of K. Then by 2.14 there exists x in K such that $x \notin S$ and $1/x \notin S$. Then $x \notin R_o$ and $1/x \notin R_o$. Since K is the quotient field of R_o, we can write $x = y_o/z_o$ with y_o and z_o in R_o and hence in M_o. Since v has center M_o in R_o, we have: $v(y_o) > 0$ and $u(z_o) > 0$. Let $x_1, \ldots, x_n$ be parameters in R_o. Arrange matters so that $v(x_1) \leq v(x_i)$ for $i = 1, \ldots, n$. Then $x_i/x_1 \in R_1$ for $i = 1, \ldots, n$ and hence $y_1 = y_o/x_1 \in R_1$ and $z_1 = z_o/x_1 \in R_1$. Thus $v(y_o) > v(y_1)$, $v(z_o) > v(z_1)$ and $x = y_o/z_o = y_1/z_1$. Proceeding in this manner, we obtain $y_i, z_i \in M_i$ such that $x = y_i/z_i$ and $v(y_o) > v(y_1) > v(y_2) > \cdots > 0$ and $v(z_o) > v(z_1) > v(z_2) > \cdots > 0$. This contradicts the discreteness of v. Therefore, S must be the valuation ring of a valuation w of K. Since R_s has a first quadratic transform R_{s+1}, it follows by the definition of quadratic transforms that $m > 1$, i.e., that v is of positive R_t-dimension whenever $t \geq s$. Now $R_v \supset R_w$ and

$$R_w \cap M_v = S \cap M_v = \left(\bigcup_{i=0}^{\infty} R_i \right) \cap M_v = \bigcup_{i=0}^{\infty} (R_i \cap M_v) = \bigcup_{i=0}^{\infty} M_i = N = M_w .$$

Therefore $v = w$, i.e., $R_v/M_v = S/N$ is an algebraic extension of R_s/M_s. Thus our assumption that the quadratic sequence $R_o < R_1 < R_2 < \cdots$ is infinite is absurd. Let R_h be the last member of this sequence. Then R_h is a regular one-dimensional local domain, i.e., R_h is the valuation ring of a real discrete valuation of K. Since $R_h \subset R_v$, we must have $R_h = R_v$. Now let $T = R_{h-1}/M_{h-1}$ and let d be the dimension of R_{h-1}. Then $d > 1$. Let $x_1, x_2, \ldots, x_d$ be a minimal basis of M_{h-1} arranged so that $v(x_1) \leq v(x_i)$ for $i = 1, 2, \ldots, d$. Let $A = R_{h-1}[x_2/x_1, x_3/x_1, \ldots, x_d/x_1]$ and $P = A \cap M_h$. As in 3.19, we may identify $A/(x_1A)$ with the polynomial ring $A^* = T[Y_2, Y_3, \ldots, Y_d]$. Let $P^* = P/(x_1A)$. As in the proof of 3.20, rank $P^* = (\dim R_h) - 1 = 0$, i.e., $P^* = (0)$. Since $R_v/M_v = R_h/M_h$ is isomorphic to the quotient field of A^*/P^*, i.e., to $T(Y_2, Y_3, \ldots, Y_d)$, we conclude that R_v/M_v is a pure transcendental extension of T of transcendence degree $d - 1 > 0$. By 2.47, R_v/M_v is finitely generated over R/M and hence T is also finitely generated over R/M; this also follows from the fact that R_{h-1} is a finite transform of R.

LEMMA 4.5. Let $R_0 < R_1 < R_2 < \cdots$ be a strictly ascending sequence of two-dimensional algebraic regular local domains with a common quotient field K, let M_i be the maximal ideal in R_i, and let $S = \bigcup_{i=0}^{\infty} R_i$. Assume that R_{i+1} is a quadratic transform of R_i for $i = 0, 1, 2, \ldots$. Then:

(1) S is the valuation ring of a valuation v^* of K such that v^* has center M_i in R_i and such that v^* is of R_i-dimension zero for each i. Furthermore:

(2) If v is any valuation of K with center M_i in R_i for $i = 0, 1, 2, \ldots$, then $v = v^*$.

PROOF. By introducing all the intermediate successive quadratic transforms between R_i and R_{i+1} for each i, we may assume without loss of generality, since this would not change S, that R_{i+1} is a first quadratic transform of R_i for $i = 0, 1, 2, \ldots$. Now suppose, if possible, that S is not the valuation ring of a valuation of K. Then by 4.1 there exists a valuation w of K having center M_i in R_i and of positive R_i-dimension for each i; and hence by 2.47 w is a prime divisor for R_0. Therefore by 4.4 the sequence $R_0 < R_1 < R_2 < \cdots$ must be finite, which is absurd. Hence $S = R_{v^*}$ where v^* is a valuation of K. Since, as in the proof of 4.1, $M_{v^*} = \bigcup_{i=0}^{\infty} M_i$, it follows that

$$R_i \cap M_{v^*} = R_i \cap \left(\bigcup_{j=0}^{\infty} M_j\right) = R_i \cap \left(\bigcup_{j=i+1}^{\infty} M_j\right) = \bigcup_{j=i+1}^{\infty} (R_i \cap M_j) = M_i ,$$

i.e., v^* has center M_i in R_i for each i. Again by 4.4 and 2.47, it follows that the R_i-dimension of v^* is zero for each i. Finally, if v is any other valuation of K with center M_i in R_i for each i, then $R_v \supset \bigcup_{i=0}^{\infty} R_i = R_{v^*}$ and $M_v \cap R_{v^*} = \bigcup_{i=0}^{\infty}(M_v \cap R_i) = \bigcup_{i=0}^{\infty} M_i = M_{v^*}$ and hence $v = v^*$.

16. <u>Uniformization in a two-dimensional regular local domain</u>. The geometric content of the following proposition is: The singularities of a curve lying on a nonsingular algebraic surface can be resolved by quadratic transformations applied to the surface.

PROPOSITION 4.6. Let (R, M) be a two-dimensional algebraic regular local ring and let K be the quotient field of R. Let P be a minimal prime ideal in R and let w be the valuation of K with $R_w = R_P$. Let v be a valuation of K composite with w and having center M in R. Let (R_n, M_n) be the n-th quadratic transform of R along v and let $P_i = R_i \cap M_v$. Then P_i is a minimal prime ideal in R_i for $i = 1, 2, \ldots$; and there exists an integer n such that for any $i \geq n$ we can choose a basis (x_i, y_i) of M_i for which $x_i R_i = P_i$, $v(y_i) = (0, 1)$, and $v(x) = (1, a)$ where a is some integer.

PROOF. Since $M_i \cap R = M$ and $P_i \cap R = R \cap M_w = P$ and since R_i is two-dimensional, P_i must be a minimal prime ideal in R_i. Now let (x, y) be a basis of M, and suppose for instance that $v(x) \leq v(y)$. Then $w(x) \leq w(y)$. Therefore $x \notin P$; for otherwise we would have $w(x) > 0$ and hence $w(y) > 0$, i.e., $x \in R \cap M_w = P$ and $y \in R \cap M_w = P$, and hence $M = P$, which is a contradiction. Let $R^* = R[y/x]$, $M^* = R^* \cap M_v$, $P^* = R^* \cap M_w$. Let $\bar{R} = R/P$, $\bar{M} = M/P$, $\bar{R}^* = R^*/P^*$, $\bar{M}^* = M^*/P^*$, $\bar{R}_i = R_i/P_i$, $\bar{M}_i = M_i/P_i$, $\bar{K} = R_w/M_w$, $\bar{v}$ = the valuation of $\bar{K}$ induced by v, and $\bar{x}$, $\bar{y}$ the residue classes modulo M_w respectively of x, y. Then $(\bar{R}_i, \bar{M}_i)$ is a one-dimensional local domain with quotient field $\bar{K}$, the real discrete valuation $\bar{v}$ of $\bar{K}$ has center $\bar{M}_i$ in $\bar{R}_i$, and $\bar{R} = \bar{R}_0 \subset \bar{R}_1 \subset \bar{R}_2 \subset \ldots$. Now $R_1 = R^*_{M^*}$, $M_1 = M^* R_1$ and $0 \neq \bar{x} \in \bar{R}$. Therefore $(\bar{x}, \bar{y})\bar{R} = \bar{M}$, $\bar{R}^* = \bar{R}[\bar{y}/\bar{x}]$, $\bar{M}^* = \bar{R}^* \cap M_{\bar{v}}$ and $\bar{R}_1 = \bar{R}^*_{\bar{M}^*}$. Hence $\bar{M}\bar{R}^* = \bar{x}\bar{R}^*$ and $\bar{M}\bar{R}_1 = \bar{x}_1\bar{R}_1$. Similarly $\bar{M}_i\bar{R}_{i+1} = z_{i+1}\bar{R}_{i+1}$ with $z_{i+1} \in \bar{R}_{i+1}$ for $i = 0, 1, 2, \ldots$.

We shall now show that $\bigcup_{i=0}^{\infty}\bar{R}_i = R_{\bar{v}}$. Given c in $R_{\bar{v}}$ we can write $c = a/b$ with $0 \neq b$, $a \in \bar{R}$. If $b \notin \bar{M}$ then $c \in \bar{R}$. Suppose that $b \in \bar{M}$. Since $\bar{v}(a) \geq \bar{v}(b)$ we must have $a \in \bar{M}$. Hence $b = b_1 z_1$, $a = a_1 z_1$ with a_1, b_1 in $\bar{R}_1$. Since $c = a_1/b_1$, again either $c \in \bar{R}_1$ or $a_1, b_1 \in \bar{M}_1$. Suppose that $c \notin \bar{R}_1$. Then $a_1 = a_2 z_2$, $b_1 = b_2 z_2$ with $a_2, b_2 \in \bar{R}_2$. Similarly if $c \notin \bar{R}_{n-1}$ then $a = a_n z_1 z_2 \ldots z_n$ and $b = b_n z_1 z_2 \ldots z_n$ with $a_i, b_i \in \bar{R}_i$. Since $\bar{M}_{n-1} = \bar{R}_{n-1} \cap M_{\bar{v}}$, we must have $\bar{v}(z_n) > 0$, and hence that $\bar{v}(b) \geq n$ where we take Z as the value group of the real discrete valuation $\bar{v}$. Since $\bar{v}$ is real discrete and since $b \neq 0$, for some n we must have $c \in \bar{R}_n$. Therefore $\bigcup_{i=0}^{\infty}\bar{R}_i = R_{\bar{v}}$. Let S_i be the integral closure of $\bar{R}_i$ in $\bar{K}$. Then by 3.18, S_i is a Dedekind domain with a finite number of prime ideals, and since the quotient ring of S_i with respect to any prime ideal is a real discrete valuation ring, we must have $S_i = \bigcap_{u \in W_i} R_u$ where W_i is a finite set of real discrete valuations of $\bar{K}$. It is clear that the valuations in W_i are exactly the valuations of $\bar{K}$ having center $\bar{M}_i$ in $\bar{R}_i$. Therefore $W_i \supset W_{i+1}$. Let $u_1, u_2, \ldots, u_h$ be the valuations in W_0 different from $\bar{v}$. Since $R_{\bar{v}}$ is a maximal subring of $\bar{K}$, we can find $a_i \in R_{\bar{v}}$ such that $a_i \notin R_{u_i}$. Since $\cup_{j=0}^{\infty}\bar{R}_j = \bar{R}_{\bar{v}}$, $a_i \in \bar{R}_{m_i}$ for some integer m_i. Let $m = \max(m_1, m_2, \ldots, m_h)$. Then $a_i \in \bar{R}_m$ for $i = 1, 2, \ldots, h$. Therefore $W_m = \{\bar{v}\}$, i.e., $R_{\bar{v}}$ is the integral closure of $\bar{R}_m$. By 3.18A, $R_{\bar{v}}$ is a finite module over $\bar{R}_m$. Hence, by the Hilbert basis theorem, we can find n such that $\bar{R}_i = \bar{R}_{\bar{v}}$ whenever $i \geq n$; from now on we shall assume that $i \geq n$. Fix $\bar{y}_i$ in $\bar{R}_i$ with $\bar{v}(\bar{y}_i) = 1$. Let y_i be an element of R_i belonging to the residue class $\bar{y}_i$. Then $v(y_i) = (0, 1)$. By 4.3, P_i is a principal ideal; let $P_i = x_iR_i$. Since $(R_i)_{P_i} = R_w$, $v(x_i) = (1, a)$. Finally, given z in M_i let $\bar{z}$ be the residue class of z modulo P_i. Then $\bar{z} \in \bar{M}_i$ and hence $\bar{z} = \bar{y}_i\bar{t}$ with $\bar{t} \in \bar{R}_i$. Fix t in R_i belonging to the residue class $\bar{t}$. Then $z - y_it \in P_i$, i.e., $z \in (x_i, y_i)R_i$. Therefore $(x_i, y_i)R_i = M_i$.

The following is the theorem referred to in the title of this section:

THEOREM 4.7. Let (R, M) be a two-dimensional regular algebraic local domain with quotient field K. Let v be a valuation of K having center M in R and R-dimension zero. Let f be a given nonzero element of R. Then there exists a quadratic transform (R^*, M^*) of R along v and a basis (x^*, y^*) of M^* such that $f = x^{*a}y^{*b}d$, where a and b are nonnegative integers, d is a unit in R^*, and where the following conditions are satisfied:

(A) If v is real of rational rank one, then $b = 0$.

(B) If v is real of rational rank two, then either $b = 0$ or $v(x^*)$ and $v(y^*)$ form an integral basis for the value group of v.

(C) If v is of rank two, then $v(x^*) = (1, h)$ and $v(y^*) = (0, 1)$ where we are writing the v-values of elements of K as lexicographically ordered pairs of integers and where h is some integer.

PROOF. First assume that v is real. Let $R_0 = R$ and $M_0 = M$. Let (x_0, y_0) be a basis of M_0 and let (R_i, M_i) be the i-th quadratic transform of R_0 along v. We shall define elements (x_i, y_i) of R_i by induction on i. Let then i be positive and assume that the definition has been made for all smaller values of i. Suppose first that $v(y_{i-1}) \geq v(x_{i-1})$. Let $S_i = R_{i-1}[y_{i-1}/x_{i-1}]$ and $N_i = M_{i-1}S_i$. Let $P_i = M_v \cap S_i$. Let z_i be the residue class modulo N_i containing y_{i-1}/x_{i-1}. Then by 3.19, $S_i/N_i = (R_{i-1}/M_{i-1})[z_i]$ and z_i is transcendental over R_{i-1}/M_{i-1}. Since P_i is a maximal ideal in S_i containing N_i, P_i/N_i must be a maximal ideal in S_i/N_i and hence $P_i/N_i = g_i(z_i)(S_i/N_i)$ where $g_i(X)$ is a monic irreducible polynomial in $(R_{i-1}/M_{i-1})[X]$. Let $G_i(X)$ be a monic polynomial in $R_{i-1}[X]$ which when reduced modulo M_{i-1} gives $g_i(X)$. We set $x_i = x_{i-1}$ and $y_i = G_i(y_{i-1}/x_{i-1})$. Secondly, if $v(y_{i-1}) < v(x_{i-1})$ then we set $x_i = y_{i-1}$ and $y_i = x_{i-1}/y_{i-1}$. Then $(x_i, y_i)R_i = M_i$.

Let $f_o = f$ and define by induction $f_i \in R_i$ by the equation $f_{i-1} = x_i^{u_i} f_i$ where u_i is a nonnegative integer and where f_i is an element in R_i prime to x_i (by 4.3, R_i is a unique factorization domain). Let $f_{i,1}, f_{i,2}, \ldots, f_{i,h_i}$ be the irreducible factors of f_i in R_i and let $w_{i,j}$ be the valuation of K whose valuation ring is the quotient ring of R_i with respect to $f_{i,j}R_i$. Let W_i be the set of valuations u of K such that u has center M_i in R_i and u is composed with $w_{i,j}$ for some j. By 4.2, W_i is a finite set. For a given u in W_i, let $P_i = M_u \cap R_{i-1}$. Since u is nontrivial, $P_i \neq (0)$. Since $x_i \notin M_u \cap R_i$ and since $x_iR_i = M_{i-1}R_i$, $P_i \neq M_{i-1}$. Therefore P_i is a minimal prime ideal in R_{i-1}. Since R_{i-1} is a unique factorization domain and since $f_{i-1} \in P_i$, we must have $P_i = f_{i-1,j}R_{i-1}$ for some j, i.e., $u \in W_{i-1}$. Thus $W_i \subset W_{i-1}$. Since by 4.5, $\cup_{i=1}^{\infty} R_i = R_v$, since v is real, and since no element of W_i is real, it follows that $\cap_{i=1}^{\infty} W_i = \emptyset$. Since W_i is finite, $W_m = \emptyset$ for some m, i.e., f_m is a unit in R_m. Hence we have $f = x_m^A y_m^B D$, where A and B are nonnegative integers and D is a unit in R_m. If either v is of rational rank one or if v is of rational rank two and $v(x_m)$ and $v(y_m)$ are rationally dependent, then the proof can be completed thus: Let $v(y_m)/v(x_m) = h_o/h_1$, where h_o and h_1 are coprime integers; by induction, let us define pairs of integers (g_i, h_{i+1}) by the equation $h_{i-1} = g_ih_i + h_{i+1}$ where $g_i \geq 0$ and $0 < h_{i+1} < h_i$; since h_o and h_1 are coprime, for some integer n we must have $h_n = 1$, let $g = g_1 + \cdots + g_n$, then it is clear that $x_m = x_{m+g}^E D_1$ and $y_m = x_{m+g}^F D_2$ where E, F are positive integers and D_1, D_2 are units in R_{m+g}. Thus we have $f = x_{m+g}^a d$ where $a = EA + FB \geq 0$ and $d = D_1D_2D =$ a unit in R_{m+g}.

Now suppose that v is of rational rank two and that $v(x_m)$ and $v(y_m)$ are rationally independent. We may then take R^* to be R_m and $x^* = x_m, y^* = y_m$. It then remains to be shown that $v(x^*)$ and $v(y^*)$ form an integral basis for the value group of v. Let $v(x^*) = p$ and $v(y^*) = q$; and suppose for instance that $p < q$. Fix a representative system k in R^* of R^*/M^* (k is not in general a field; note that we take zero as the representative of M^*/M^*). Let z be an arbitrary nonzero element of R^* and let $v(z) = r$. Fix an integer n so that $r < np$. Then we can find a polynomial $H(X, Y) = \Sigma_{i+j \leq n} H_{ij} X^i Y^j$ of degree at most n with coefficients H_{ij} in k such that $z^* = z - H(x^*, y^*) \in M^{*n}$. Since $v(u) > r$ for any $u \in M^{*n}$, we must have that $v(z^*) > r$ and hence that $r = v(z) = v(H(x^*, y^*))$. Since p and q are rationally independent and since $v(H_{ij}) = 0$ whenever $H_{ij} \neq 0$, we can find $H_{st} \neq 0$ such that $v(H_{st}x^{*s}y^{*t}) < v(H_{ij}x^{*i}y^{*j})$ whenever $H_{ij} \neq 0$ and whenever either $i \neq s$ or $j \neq t$. Therefore $r = v(z) = v(H(x^*, y^*)) = v(H_{st}x^{*s}y^{*t}) = sp + tq$. Therefore for any nonzero element z_1 of K, we must have $v(z_1) = s_1p + t_1q$ where s_1 and t_1 are integers. Thus we have shown that p, q is an integral basis of the value group of v. This completes the proof of (A) and (B).

To prove (C), assume v is of rank two, and that we are writing the values of elements of K as lexicographically ordered pairs of integers. By 4.6, we can find a quadratic transform $(\bar{R}, \bar{M})$ of R along v and a basis $\bar{x}, \bar{y}$ of $\bar{M}$ such that $v(\bar{x}) = (1, p)$ and $v(\bar{y}) = (0, 1)$ where p is some integer. Let $(\bar{R}_i, \bar{M}_i)$ be the i-th quadratic transform of $\bar{R}$ along v. Let $\bar{x}_i = \bar{x}/\bar{y}_i$ and $\bar{y}_i = \bar{y}$. Since $v(\bar{x}) > iv(\bar{y})$ for any integer i, it follows that $(\bar{x}_i, \bar{y}_i)\bar{R}_i = \bar{M}_i$. Let $f = \bar{x}^a g_0$ where a is a nonnegative integer and where g_0 is an element of $\bar{R}$ prime to $\bar{x}$. Then $v(g_0) = (0, t)$ where t is some nonnegative integer. Define g_i by induction by the equation $g_{i-1} = \bar{y}_i^{e_i} g_i$ where e_i is a nonnegative integer and g_i is an element of $\bar{R}_i$ prime to $\bar{y}_i$. Since $\bar{M}_{i-1}\bar{R}_i = \bar{y}_i\bar{R}_i$, we have that $e_i > 0$ whenever g_{i-1} is a nonunit in $\bar{R}_{i-1}$ (i.e., whenever $g_{i-1} \in \bar{M}_{i-1}$). Therefore, if g_{i-1} is a nonunit in $\bar{R}_{i-1}$ for $i = 1, 2, \ldots, n$, then $v(g_0) \geqq (0, n)$. Hence for some integer $m \leqq t$, g_m must be a unit in $\bar{R}_m$. Hence $f = x^{*a}y^{*b}d$ where b is a nonnegative integer, d is a unit in $R^* = \bar{R}_m$, $x^* = \bar{x}_m$, and $y^* = \bar{y}_m$.

THEOREM 4.8. Let L/k be a two-dimensional algebraic function field, K a finite algebraic extension of L, v a zero dimensional valuation of K/k, and (R, M) a regular algebraic local domain with quotient field K and ground field k such that v has center M in R and dim R = k-rank R = 2. Then for some quadratic transform R* of R along v there exists an algebraic local domain S* in L with ground field k lying below R*.

PROOF. Let q_1, q_2 be a transcendence basis of L/k. Replacing q_1 by $1/q_1$ if necessary, we may assume that $q_1, q_2 \in R_v$. Replacing R by a quadratic transform along v we may assume that $q_1, q_2 \in R$ (see 4.5). Let $S = R \cap L$, $Q = M \cap L$ and $P = QR$. If e is an element of L which is integral over S, then e is integral over R, i.e., e is in R and hence e is in $R \cap L = S$. Therefore S is integrally closed in L. Since $q_1, q_2 \in R \cap L = S$, L is algebraic over S and hence L must be the quotient field of S.

Let w be the L-restriction of v. Fix an arbitrary element z_1 in L with $w(z_1) > 0$. Now we choose an element z_2 in L as follows: if the rational rank of w is one, then let $z_2 = 1$ and if the rational rank of w is two, then let z_2 be an element of M_w such that $w(z_1)$ and $w(z_2)$ are rationally independent. Now let $z = z_1 z_2$. By 4.5 and 4.7, there exists a quadratic transform (R^*, M^*) of R along v such that $z_1, z_2 \in R^*$ and $z = x^a y^b d$ where (x, y) is a basis of M^*, a and b are nonnegative integers, d is a unit in R^* and where we have either that $b = 0$ and $a > 0$ or that $v(x)$ and $v(y)$ are rationally independent and constitute an integral basis for the value group of v. We let $S^* = R^* \cap L$, $Q^* = M^* \cap L$ and $P^* = Q^*R^*$ and we proceed to show that in either case P^* is primary for M^*; and that will, then, in view of 3.12, complete the proof.

CASE 1: $a > 0$ and $b = 0$. Now L is the quotient field of $S^* = R^* \cap L$ and y is algebraic over L. Hence for some nonzero element e of S^*, ey is integral over S^*. Let $(ey)^n + f_{n-1}(ey)^{n-1} + \cdots + f_1(ey) + f_o = 0$ be the minimal equation of ey over L. Then $f_o, f_1, \ldots, f_n$ belong to S^* where we take $f_n = 1$. Let $g_i = f_i e^i$. Then $g_o, g_1, \ldots, g_n$ belong to S^* and we have $g_n y^n + \cdots + g_o = 0$. The minimality of the equation implies that $g_o \neq 0$. Thus $g_o = -y(g_n y^{n-1} + g_{n-1} y^{n-2} + \cdots + g_1)$ is a nonzero element of $S^* = R^* \cap L$ which is divisible by y in R^*. Let $g_o = x^u h^*$ where u is a nonnegative integer and h^* is a nonzero element of R^* which is not divisible by x. Let $h = g_o^a / z^u$. Then h is a nonzero element of $R^* \cap L = S^*$ which is not divisible by x in R^*, but is divisible by y in R^*. Since h is in $yR^* \subset M^*$ and h is in L, we have that h is in $M^* \cap L = Q^*$. Suppose if possible that P^* is not primary for M^*. Then $P^* \subset tR^*$ where t is a nonzero nonunit in R^* (since every rank one prime ideal in R^* is principal). Now $z = x^a d \in P^* \subset tR^*$ and hence $tR^* = x^{a^*}R^*$, where a^* is a positive integer $(a^* \leq a)$. But $h \in P \subset tR^* = x^{a^*}R^*$ is a nonunit in R^* which is not divisible by x. This is a contradiction. Hence P^* is primary for M^*.

CASE 2: $a > 0$ and $b > 0$. In this case, w is of rational rank two and hence $v(z_1)$ and $v(z_2)$ are rationally independent. Since $z_1 z_2 = z = x^a y^b d$ and since $z_1, z_2 \in R^*$, we must have $z_1 = x^{a_1} y^{b_1} d_1$ and $z_2 = x^{a_2} y^{b_2} d_2$ where a_1, b_1, a_2, b_2 are nonnegative integers and d_1 and d_2 are units in R^*. Let $z_3 = z_1^{a_2} z_2^{-a_1}$ and $z_4 = z_1^{b_2} z_2^{-b_1}$. Then we have $z_3 = y^{b_3} d_3$ and $z_4 = x^{a_4} d_4$, where b_3 and a_4 are integers and d_3 and d_4 are units in R^*. Let D denote the determinant with first row (a_1, b_1) and second row (a_2, b_2). Then we have

$$z_1^D = z_3^{-b_1} z_4^{a_1} \quad \text{and} \quad z_2^D = z_3^{-b_2} z_4^{a_2} \quad .$$

Since $v(z_1)$ and $v(z_2)$ are rationally independent, we must have $D \neq 0$ and hence in view of the above two equations we conclude that $v(z_3)$ and $v(z_4)$ are also rationally independent and hence $v(z_3) \neq 0 \neq v(z_4)$. Therefore $b_3 \neq 0 \neq a_4$. Now we define elements z_1^* and z_2^* as follows: if $b_3 > 0$ then let $z_1^* = z_3$ and if $b_3 < 0$ then let $z_1^* = z_3^{-1}$; if $a_4 > 0$ then let $z_2^* = z_4$ and if $a_4 < 0$ then let $z_2^* = z_4^{-1}$. Then

$$z_1^* = y^{b^*}d_1^* \quad \text{and} \quad z_2^* = x^{a^*}d_2^* \quad ,$$

where a^* and b^* are positive integers and d_1^* and d_2^* are units in R^*. Since z_1^* and z_2^* belong to $M^* \cap L = Q^*$, we conclude that P^* is primary for M^*.

17. Uniformization for a two-dimensional algebraic function field. Let K/k be a two-dimensional algebraic function field and let v be a valuation of K/k. The theorem of local uniformization asserts that v can be uniformized, i.e., there exists a regular algebraic local domain (R, M) with quotient field K and ground field k such that v has center M in R and k-rank of R = k-dimension of v. This theorem was proved by Zariski for fields of zero characteristic in [Z3] and [Z4] and for perfect fields of nonzero characteristic by the author in [A2] and [A6]. As an application of ramification theory, we give in this section a simple proof for the case when k is algebraically closed and of characteristic zero. If k-dimension of v is one, we fix $a_1, \ldots, a_n$ in R_v such that K is the quotient field of $A = k[a_1, \ldots, a_n]$, A is normal, and the residue class of a_1 in D_v is transcendental over k, we let $P = A \cap M_v$, $R = A_P$ and $M = PR$, then (R, M) is regular of dimension one (see 3.15) and v has center M in R. Thus it is enough to consider the case when v is zero dimensional (i.e., k-dimension $v = 0$).

THEOREM 4.9. (Theorem of uniformization for K/k). Let K be a two-dimensional algebraic function field over an algebraically closed ground field k of characteristic zero. Let v be a zero dimensional valuation of K/k. Then v can be uniformized.

PROOF. By 2.49, $r(v) = 1$ or $r(v) = 2$ and v is an integral direct sum. If $r(v) = 2$, choose x_1, x_2 in K such that $v(x_1), v(x_2)$ is an integral basis for the value group of v; then by 2.46, (x_1, x_2) is a transcendence basis of K/k. If $r(v) = 1$, then let (x_1, x_2) be an arbitrary transcendence basis of K/k. Replacing x_1 by x_1^{-1} if necessary we may assume that $v(x_1) \geq 0$; since $D_v = k$, subtracting suitable elements of k from x_1, x_2 we may assume that $v(x_i) > 0$ for $i = 1, 2$. Let $K_1 = k(x_1, x_2)$, and let (R_1, M_1) be the quotient ring of $k[x_1, x_2]$ with respect to the prime ideal generated by (x_1, x_2). Then v has center M_1 in R_1 and $\dim R_1 = 2$ and $(x_1, x_2)R_1 = M_1$, so that R_1 is regular.

Let K^* be a galois extension of K_1 containing K and let v^* be an extension of v to K^* and let v_1 be the K_1-restriction of v. Observe that if $r(v) = 2$, then $S_{v_1} = S_v$. Let $v^* = v_1^*, v_2^*, \ldots, v_n^*$ be all the K^*-extensions of v_1. Then $T = \cap_{i=1}^n R_{v_i^*}$ is the integral closure of R_{v_1} in K^*, and $H_i = T \cap M_{v_i^*}$ for $i = 1, \ldots, n$ are the maximal ideals in T. Hence by (1.3) there exists $u \in T$ such that $u \in H_1$ and $u \notin H_i$ for $i = 2, \ldots, n$. Let $u^m + a_1u^{m-1} + \ldots + a_m = 0$, with $a_i \in R_{v_1}$ be the equation of integral dependence of u over R_{v_1}. In view of 4.5, by replacing R_1 with a quadratic transform along v_1 we may assume that $a_i \in R_1$ for $i = 1, \ldots, m$. Let S be the integral closure of R_1 in K^*, let $P_1 = S \cap M_{v^*} = S \cap H_1$. Then $u \in S \cap H_1 = P_1$ and $u \notin H_i$ for $i = 2, \ldots, n$ implies that $P_1 \neq S \cap H_i$ for $i = 2, \ldots, n$. Let $g \in G(K^*/K_1)$ (= the galois group of K^* over K_1). Then $gP_1 = P_1$ implies that $gH_1 = H_1$. Therefore $G^s(H_1/M_{v_1}) \supset G^s(P_1^*/M_1)$. By 1.50, $G^s(H_1/M_{v_1}) \subset G^s(P_1^*/M_1)$. Hence $G^s(P_1^*/M_1) = G^s(H_1/M_{v_1}) = G^s(v^*/v)$. Let $R^* = S_{P_1}$ and $M^* = P_1R^*$. Then (R^*, M^*) is the local ring in K^* lying above R_1 such that v^* has center M^* in R^* and we have $G^s(R^*/R) = G^s(v^*/v)$. Let $R = R^* \cap K$ and $M = M^* \cap K$. Let $K_1^s = F^s(R^*/R_1)$ and $K^s = F^s(R^*/R)$. Then by 1.49, K^s is the compositum of K_1^s and K. Since $K = D_v$ is algebraically closed, by 1.48, we have $G^i(v^*/v) = G^s(v^*/v)$; and by 2.43, $G^i(v^*/v)$ is abelian. Therefore $G(K^*/K_1^s) = G^i(v^*/v)$ and this is abelian. Hence K^s is galois over K_1^s and $G(K^s/K_1^s)$ is abelian. Hence we can find a sequence of fields $K_1^s = L_o \subset L_1 \subset \ldots \subset L_t = K^s$ such that L_i is a cyclic extension of L_{i-1} of prime degree for $i = 1, \ldots, t$. Let w_i be the L_i-restriction of v^*. Since v^* is the only K^*-extension of w_o, w_i is the only extension of w_{i-1} to L_i for $i = 1, \ldots, t$. Let $S_i = R^* \cap L_i$ and $N_i = M^* \cap L_i$. By 1.47, $M_1S_o = N_o$ and hence the regularity of R_1 implies the regularity of S_o. Thus

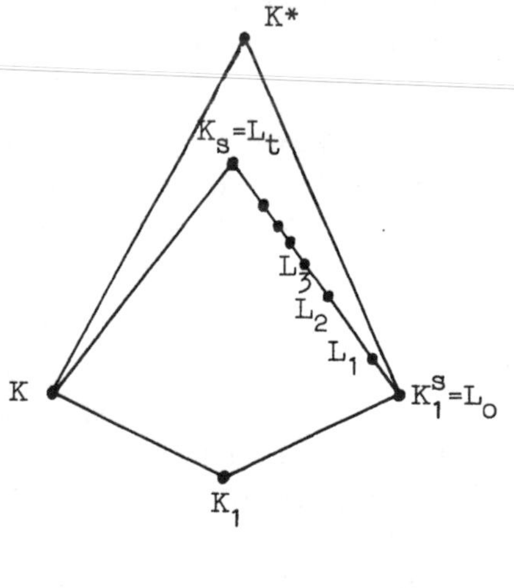

w_o has been uniformized. Now we anticipate the following

PROPOSITION 4.10. Let L be a two-dimensional algebraic function field over an algebraically closed ground field k of characteristic zero. Let L^* be a cyclic extension of L of prime degree p $(p > 1)$. Let w be a zero dimensional rational valuation of L/k having a unique extension w^* to L^*. Assume that w can be uniformized. Then w^* can be uniformized.

If $r(v) = 1$, then applying this proposition successively to the extensions L_1/L_o, L_2/L_1, ..., L_t/L_{t-1} we conclude that w_t can be uniformized. If $r(v) = 2$ then $S_{v_1} = S_v$ and by 2.40, $S_v = S_{w_t}$ and $S_{v_1} = S_{w_o}$ so that $S_{w_t} = S_{w_o}$ and since w_t is the only extension of w_o to L_t and $D_{w_o} = k = D_{w_t}$, by 1.46 and 1.48 we have that $L_o = F^i(w_t/w_o)$ and hence by 2.43, $1 = [S_{w_t} : S_{w_o}] = [L_t : L_o]$, i.e., $L_t = L_o$; therefore $w_t = w_o$ can be uniformized. Let $v^s = w_t$, $R^s = S_t$ and $M^s = N_t$. Observe that by 1.49, $K^s = F^s(v^*/v)$. Since v^s can be uniformized, in view of 4.5 we can find a two-dimensional regular local domain $(\bar{R}^s, \bar{M}^s)$ with quotient field K^s and ground field k such that v^s has center $\bar{M}^s$ in $\bar{R}^s$ and $\bar{R}^s \supset R^s$. In view of 4.8, we may assume that there exists a local ring $(\bar{R}, \bar{M})$ in K lying below $\bar{R}^s$. Let $(\bar{R}^*, \bar{M}^*)$ be the local ring in K^* lying above $\bar{R}^s$, for which v^* has center $\bar{M}^*$ in $\bar{R}^*$. By 1.50, $F^s(v^*/v) \supset F^s(\bar{R}^*/\bar{R}) \supset F^s(R^*/R)$. Since $F^s(v^*/v) = F^s(R^*/R) = K^s$, we conclude that $K^s = F^s(\bar{R}^*/\bar{R})$. Hence, by 1.47, $\bar{M}\bar{R}^* = \bar{M}^*$ and $\bar{R}^*/\bar{M}^* = \bar{R}/\bar{M}$. Therefore, in view of 3.18B, the regularity of $\bar{R}^*$ implies the regularity of $\bar{R}$. Since v has center $\bar{M}$ in $\bar{R}$, we have uniformized v.

Thus the proof of Proposition 4.10 will complete the proof of Theorem 4.9.

PROOF of Proposition 4.10. Let (R, M) be a two-dimensional regular algebraic local domain with quotient field K and ground field k such that w has center M in R. Let (R^*, M^*) be the local ring in L^* lying above R (since v has a unique extension to L^*, R^* is unique). Since L^*/L is cyclic of prime degree p and since k is algebraically closed and of characteristic zero, there exists a primitive element z of L^*/L whose minimal polynomial is of the form

$$f(Z) = Z^p - u, \quad \text{with } u \in L, \ (u \neq 0).$$

Multiplying z by a suitable element in R we may assume that $u \in R$. In view of 4.8, we may assume that

$$u = x^a d \quad ,$$

where (x, y) is a basis of M, d is a unit in R and a is a non-negative integer. Suppose if possible that $a \equiv 0(p)$. Then replacing z by $zx^{-a/p}$, we may assume that $u = d$; this would imply that $D_{L^*/L}(1, z, \ldots, z^{p-1}) = (-d)^{p-1}$ = a unit in R so that $D_{L^*/L}(L^*/R) = R$ and in view of 1.42 this is a contradiction since $[L^* : L] = h > 1$ and R^* is the only local ring in L^* above R and $[(R^*/M^*) : (R/M)]_s = [k : k] = 1$.

Therefore $a \not\equiv 0(p)$. Hence for some positive integer h we have $ah \equiv 1(p)$. Replacing z by $z^h x^{(1-ah)/p}$ we may assume that

$$u = (x^a d)^h [x^{(1-ah)/p}]^p = x^{ah} d^h x^{1-ah} = xd^h \quad .$$

Since d^h is unit in R we may replace x by xd^h and assume that $u = x$, i.e.,

$$z^p = x \; .$$

Let $S = R[z]$ and $N = S \cap M^*$. Let $S^* = S_N$ and $N^* = NS^*$. Then (S^*, N^*) is an algebraic local domain with quotient field L^* and ground field k, and the zero dimensional valuation w^* of L^*/k has center N^* in S^*. Therefore $\dim S^* = 2$. Since $N^* = (x, y, z)S^* = (y, z)S^*$, S^* is regular and hence normal. Since $R \subset S^* \subset R^*$ and since R^* is the integral closure of R in L^*, we must have $R^* = S^*$.

REMARK 4.11. Let K be an n-dimensional algebraic function field over an algebraically closed ground field k and let v be a zero dimensional valuation of K/k. It can be shown that "to uniformize v" is equivalent "to finding a transcendence basis $x_1, \ldots, x_n$ of K/k with $v(x_1) > 0$ and a $k(x_1, \ldots, x_n)$-isomorphism t of K into the formal power series field $K((x_1, \ldots, x_n))$".

For suppose v can be uniformized, and let (R, M) be an n-dimensional regular algebraic local ring with quotient field K and ground field k such that v has center M in R and let $x_1, \ldots, x_n$ be a system of parameters in R. Then it can be shown that $x_1, \ldots, x_n$ is a transcendence basis of K/k ($v(x_1) > 0$ is obvious since v has center M in R) and that there exists a unique $k(x_1, \ldots, x_n)$-isomorphism t of K into $k((x_1, \ldots, x_n))$.

Conversely, let $x_1, \ldots, x_n$ be a transcendence basis of K/k with $v(x_1) > 0$ and let t be a $k(x_1, \ldots, x_n)$-isomorphism of K into $k((x_1, \ldots, x_n))$. Let R_1 be the quotient ring of $k[x_1, \ldots, x_n]$ with respect to the prime ideal generated by $x_1, \ldots, x_n$ and let (R, M) be the local ring in K lying above R_1 such that v has center M in R. Then it can be shown that R is regular n dimensional and $R = \{y \in K \mid t(y) \in k[[x_1, \ldots, x_n]]\}$.

CHAPTER V: VARIETIES AND TRANSFORMATIONS

Recall that a local domain (R, M) (respectively, a valuation v) is said to have center P in A, where A is a domain and P is a prime ideal in A, if $A \subset R$ and $M \cap A = P$ (respectively, $A \subset R_v$ and $M_v \cap A = P$); if (S, N) is a local domain such that R (respectively v) has center N in S then we shall say that R (respectively v) <u>has center in</u> S, or that R is <u>centered in</u> S, or that S is <u>at the center of</u> R. Also recall that a local domain S is said to be a specialization of R ($R \longrightarrow S$) if R is the quotient ring of S with respect to a nonzero prime ideal.

For a field K and a subfield k we define

$\Omega(K)$ = the <u>Riemann manifold</u> of K
= the set of all valuations of K.

$\Omega(K/k)$ = the <u>Riemann manifold</u> of K/k
= the set of all valuations of K/k.

Now let K be a field and let V and W be two sets of local domains with quotient field K. We shall say that V is <u>dominated by</u> W if every member of V is at the center of some member of W and V will be said to be <u>compatible</u> with W, if no two distinct members of V are at the center of a same member of W. If V is dominated by and compatible with $\Omega(K)$ then we shall say that V is a <u>variety</u>, that V is a <u>model</u> of K and K is the <u>function field of</u> V; by the <u>dimension</u> of V will be meant the maximum of (dim R - dim S) taken over all pairs R, S of members of V. A variety V will be said to be respectively <u>noetherian</u>, <u>locally algebraic</u>, or <u>normal</u> if each member of V is respectively noetherian, algebraic or normal. A variety will be said to be <u>closed</u> if with each member it contains all its quotient rings with respect to prime ideals: observe that a union of closed models of a function field K is a closed variety with function field K. A variety V will be said to be <u>complete</u> if each valuation of the function field of V has center in some member of V. If k is a subfield of K and if each member of V contains k then V will be said to be a <u>model</u> of K/k and K/k will be said to be a function field of V/k, V will be said to be complete over k if every valuation of K/k has center in a member of V.

Now let K/k be a n-dimensional algebraic function field and let A be a domain with quotient field K and finitely generated over k, and let $V(A)$ be the set of all quotient rings of A; then it is clear that $V(A)$ is a closed locally algebraic model of K/k; we shall say that $V(A)$ is an affine variety. If a model V of K/k is the union of a certain set of affine models of K/k, then we shall say that V is an algebraic variety with function field K/k, or function field K and ground field k; observe that $\dim V = n(= \dim K/k)$. If a model V of K/k is the union of a finite number of affine models of K/k then we shall say that V is a finitely generated algebraic variety. We remark that the "abstract varieties" introduced by A. Weil in his book (Foundations of Algebraic Geometry) are in our language exactly the "finitely generated algebraic varieties with an algebraically closed ground field". Now let $A_o, A_1, \ldots, A_m$ be subdomains of K with quotient field K and finitely generated over k; we shall say that $A_1, \ldots, A_m$ are projectively related over k if there exist elements $x_1, \ldots, x_m$ in K such that $A_o = k[x_1, \ldots, x_m]$ and $A_i = k[x_1/x_i, x_2/x_i, \ldots, x_{i-1}/x_i, 1/x_i, x_{i+1}/x_i, \ldots, x_m/x_i]$ for $i = 1, \ldots, m$; it can easily be shown that if $A_o, \ldots, A_m$ are projectively related, then $\cup_{i=0}^{m} V(A_i)$ is a variety complete over k, and hence a finitely generated algebraic model of K/k. If for a given model V of K/k there exist subdomains $A_o, \ldots, A_m$ of K which are projectively related over k such that $V = \cup_{i=0}^{m} V(A_i)$, then we shall say that V is a projective variety and that A_o is an affine coordinate ring of V; it can at once be shown that $A_1, \ldots, A_m$ are also affine coordinate rings of V.

18. Projective varieties. Let V be a projective variety of dimension n and function field K/k; an n-dimensional member of V will be called a point of V. Given a subset W of V, W is called an irreducible subvariety of V if there exists $R \in W$ such that W is the totality of the specialization of R in V, observe that if R exists then it is unique and we call it the general point of W and we set $\dim W = n - \dim R$. A subset of V is called a subvariety of V if it is the union of a finite number of irreducible subvarieties of V; observe that if W is a subvariety of V then there are only a finite number of irreducible subvarieties $W_1, \ldots, W_t$ of V such that $W_i \subset W$ and there are no irreducible subvarieties of V between W_i and W. We shall say that $W_1, \ldots, W_t$ are the irreducible components of W and we set $\dim W = \max(\dim W_1, \ldots, \dim W_t)$. Let K^* be a finite algebraic extension of K and let V^* be the set of local rings in K^* lying above

the various members of V; then it is clear that V^* is finitely generated complete algebraic model of K/k; it is proved in [Z7] that V^* is indeed a projective model of K/k. We shall call V^* the K^*-normalization of V, the totality of members R of V such that the integral closure of R is ramified over R will be called the branch locus of V for the extension K^*/K and will be denoted by $D(K^*, V)$. The totality of nonregular members of V will be called the singular locus of V, V will be said to be nonsingular or singular according as the singular locus of V is empty or nonempty.

PROPOSITION 5.1. The singular locus S of a projective variety V is a proper subvariety of V.

PROOF. See [Z2].

COROLLARY 5.2. If V is normal, then $\dim S \leq \dim V - 2$. Follows from 5.1 and 3.15.

PROPOSITION 5.2. Let V be a projective normal model of K/k and let K^* be a finite algebraic extension of K. If K^*/K is inseparable, then $D(K^*, V) = V$ and if K^*/K is separable, then $D(K^*, V)$ is a proper subvariety of V.

PROOF. Follows from the results of Section 6. See also Lemma 3 of [A3].

PROPOSITION 5.3. Let V be a nonsingular projective model K/k, let K^* be a finite separable extension of K and assume that k is algebraically closed. Then each irreducible component of $D(K^*, V)$ is of dimension $(\dim V - 1)$.

PROOF. [Z8].

Now let K/k be an algebraic function field and let K_1 and K_2 be fields between k and K. Let V_1 and V_2 be projective models of K_1/k and K_2/k, respectively; let T be the ordered pair (V_1, V_2). Let $R_1 \in V_1$ and $R_2 \in V_2$. We say that R_2 corresponds to R_1 under T if there exists a valuation of K/k having center in R_1 and R_2; for $R_1 \in V_1$ we denote by $T(R_1)$ the set of all members of V_2 which correspond to specializations of R_1; it can then be shown that $V_2(R_1)$ is a subvariety of V_2.

Now assume that $K_1 = K_2 = K$. Then $R_2 \in V_2$ is said to be fundamental for T if there exists $R_1 \in V_1$ to which R_2 corresponds under T and $R_1 \not\subset R_2$; if R_2 is not fundamental for T, then R_2 is said to be regular for T.

LEMMA 5.4. Let V_1 and V_2 be projective models of K/k. Then $R_2 \in V_2$ is regular for T if and only if there is a unique $R_1 \in V_1$ to which R_2 corresponds under T and this unique R_1 is contained in R_2.

PROOF. Exercise.

COROLLARY 5.5. V_2 dominates V_1 if and only if each member of V_2 is regular for (V_1, V_2).

PROPOSITION 5.6. Let V_1 and V_2 be normal projective models of K/k. Then the set of fundamental elements of (V_1, V_2) is a subvariety of V_2 of dimension $\leq (\dim V_2 - 2)$.

PROOF. See [Z1].

LEMMA 5.7. Given a finite number of projective models $V_1, \ldots, V_t$ of K/k, let $V_1 \oplus \cdots \oplus V_t$ denote the set of all the rings of the type $k[A_1, \ldots, A_t]$ where A_i is an affine coordinate ring of V_i, and let $V_1 + \cdots + V_t$ be the set of all the quotient rings of all the members of $V_1 \oplus \cdots \oplus V_t$. Then

(1) $V_1 + \cdots + V_t$ is a projective model of K/k, and

(2) $V_1 + \cdots + V_t$ dominates V_i for $i = 1, \ldots, t$.

PROOF. (1) is proved in [Z5] and (2) is obvious.

DEFINITION 5.8. $V_1 + \cdots + V_t$ is called the <u>join</u> of $V_1, \ldots, V_t$.

DEFINITION 5.9. Let (R, M) be a normal algebraic local domain with quotient field K and ground field k such that $\dim R = k\text{-rank } R$. Let v be a zero dimensional valuation of K/k having center M in R. Since M is finitely generated, there exists $0 \neq x \in M$ such that $v(y) \geq v(x) > 0$ for all $y \in M$; let B be the ring generated over R by all elements of the type m/x with $m \in M$, let A be the integral closure of B in K, let $P = A \cap M_v$, $S = A_P$ and $M = PS$. Then (S, N) is called the <u>immediate</u> (or <u>first</u>) <u>quadratic transform</u> of R along v. If R_i is the immediate quadratic transform of R_{i-1} along v where $R_0 = R$, we shall say that R_n is the n-th quadratic transform of R along v. Observe that if R is regular, then this definition coincides with the definition given in Section 14, and hence there is no ambiguity involved.

PROPOSITION 5.10. Let V be the projective model of K/k and let P be a regular point of V. Let $V^* = (V - P) \cup$ (the set of all the immediate quadratic transforms of P). Then

(1) V^* is a projective model of K/k,

(2) V^* dominates V and P is the only fundamental of (V^*, V),

(3) If V is normal (respectively, nonsingular), then V^* is normal (respectively nonsingular).

PROOF. (1) is proved in [Z5] and (2) and (3) are obvious.

PROPOSITION 5.11. Let R_0 be a normal algebraic two-dimensional local domain with quotient field K and ground field k such that k-rank R = 2. Let $R_0 = R$ and let (R_i, M_i) be an immediate quadratic transform of R_{i-1}. Let $S = \cup_{i=0}^{\infty} R_i$. Then

(1) S is the valuation ring of a zero-dimensional valuation v* of K/k, and

(2) If v is any valuation of K/k having center M_i in R_i for i = 0, 1, ..., then v = v* (and hence R_i is the i-th quadratic transform of R_0 along v).

PROOF. Entirely similar to the proof of 4.5.

DEFINITION 5.12. In the notation of 5.10 V* is called the quadratic transform of V with center P.

DEFINITION 5.13. Let K/k be an algebraic function field. By $\Omega_0(K/k)$ we shall denote the set of all zero-dimensional valuations of K/k. For a subset W of a projective model V of K/k, by $\Omega(W)$ we shall denote the set of all zero-dimensional valuations of K/k having center in some member of W. Let $V_1, \ldots, V_t$ be a finite number of projective models of K/k and let G be a subset of $\Omega_0(K/k)$; if for each $v \in G$ there exists regular P in $V_1 \cup \ldots \cup V_t$ at the center of v, then we shall say that $(V_1, \ldots, V_t)$ is a <u>resolving system</u> of G. Observe that a projective model V of K/k is nonsingular if and only if V is a resolving system of $\Omega_0(K/k)$.

LEMMA 5.14. Let K/k be an algebraic function field and let R be an algebraic local domain with quotient field K and ground field k. Then R is contained in a projective model of K/k.

PROOF. Obvious.

COROLLARY 5.15. Let v be a zero-dimensional valuation of a two-dimensional algebraic function field K over a perfect ground field k. Then there exists a projective model V of K/k on which v is centered in a regular point.

For algebraically closed characteristic zero ground field, it follows from 4.9; for the general case, see [A2] and [A6].

REMARK 5.16. In the remaining part of this chapter, we are going to copy the last part of [Z5] in our language.

LEMMA 5.17. Let V be a projective model of an algebraic function field K/k and let W be a subvariety of V. Assume that for each point P in W there exists a resolving system for $\Omega(P)$. Then there exists a resolving system for $\Omega(W)$.

PROOF. Let $\dim W = s$. We apply induction to s; for $s = 0$, W consists of a finite number of points and hence the lemma is trivial, so assume $s > 0$ and that the lemma is true for all smaller values of s. Let $W_1, \ldots, W_t$ be the s-dimensional components of W and fix a point P_i in W_i and let $V_1, \ldots, V_h$ be a resolving system for $\Omega(P_1, \ldots, P_t)$. Let $V^* = V + V_1 + \cdots + V_h$. The elements of V^* which correspond to non-regular elements of V_i under (V_i, V^*) form a subvariety W_i^* of V^*. Let $W^* = W_1^* \cap \ldots \cap W_h^*$. The elements of W which correspond to elements of W^* under (V^*, V) is a subvariety W_1 of V contained in W such that $\dim W_1 < s$, since $P_i \notin W_1$. It is clear that $(V_1, \ldots, V_h)$ is also a resolving system of $\Omega(W - W_1)$. By the induction hypothesis $\Omega(W_1)$ has a resolving system and this together with $V_1, \ldots, V_h$ is a resolving system of $\Omega(W)$.

PROPOSITION 5.18. Let K be a two-dimensional algebraic function field over a perfect ground field k. Then there exists a resolving system for $\Omega_o(K/k)$.

PROOF. Assume the contrary. Let V be a normal projective model of K/k. Then by 5.17 there exists a point P in V such that $\Omega(P)$ has no resolving system. Let V_1 be the quadratic transform of V with center P and let $W_1 = (V, V_1)(P)$. Then $\Omega(P) = \Omega(W_1)$ and hence by 5.17 there exists a point P_1 in W_1, i.e., a first quadratic transform of P such that $\Omega(P_1)$ has no resolving system. By repeated application of this argument, there exists a point P_i on a projective model V_i of K/k such that P_{i+1} is a first quadratic transform of P_i and such that $\Omega(P_i)$ has no resolving system. By 5.11, $\bigcup_{i=1}^{\infty} P_i = R_v$ where v is a zero-dimensional valuation of K/k. By 5.15 there exists a projective model V^* of K/k on which the point P^* at the center of v is regular. Now $P^* \subset R_v$ and hence $P^* = P_n$ for some n, and hence V^* is a resolving system for $\Omega(P_n)$. This is a contradiction.

19. Resolution of singularities of algebraic surfaces.

THEOREM 5.19. Let K be a two-dimensional algebraic function field over a perfect ground field k. Then K/k possesses a nonsingular projective model.

PROOF. In view of 5.18, it is enough to prove that if for a subset Ω of $\Omega_o(K/k)$ there exists a resolving system of two models V and V', then Ω has a resolving system consisting of a single model. Replacing V and V' by their K-normalization, if necessary, we may assume that V and V' are normal. By 5.6, there are only a finite number of points on V which are fundamental for (V', V). Let V_1 be the model obtained from V by applying successive quadratic transformations with centers at the fundamental points. If (V', V_1) has fundamental points on V_1, then obtain V_2 from V_1 as V_1 is obtained from V and so on. By 5.11, it follows that for some n, (V', V_n) has no fundamental points on V_n. Since V', V_n is also resolving system for Ω, we may assume to begin with that (V', V) has no fundamental points on V.

Now, by the above described process, we eliminate the fundamental points of (V, V'), but this time only those points which are regular. Let V'' be the transform of V' thus obtained. Let $V^* = V + V''$. We assert that V^* is a resolving system for Ω. For v in Ω let P, P', P'', P^* be the points respectively of V, V', V'', V^* centered in v. We have to show that P^* is regular.

CASE 1. <u>Assume that P' is non-regular</u>. Since V, V' is a resolving system of Ω, P is regular. Since V dominates V', $P \supset P'$. Since we have not disturbed the nonsingular points of V', we have $P'' = P'$, i.e., $P \supset P''$. Hence $P^* = P$ and P^* is regular.

CASE 2. <u>Assume that P' is regular</u>. Then P'' is regular and nonfundamental for (V, V'') so that $P'' \supset P$. Hence $P^* = P''$ and P^* is regular.

BIBLIOGRAPHY

<u>S. Abhyankar</u>

[A1] Splitting of valuations in extensions of local domains II, Proc. Nat. Acad. Sci. U.S.A., vol. 41 (1955), pp. 220-223.

[A2] Local uniformization on algebraic surfaces over ground fields of characteristic $p \neq 0$, Annals of Math., vol. 63 (1956), pp. 491-526.

[A3] On the ramification of algebraic functions, Amer. Jour. of Math., vol. 77 (1955), pp. 575-592.

[A4] On the valuations centered in a local domain, Amer. Jour. of Math., vol. 78 (1956), pp. 321-348.

[A5] Simultaneous resolution for algebraic surfaces, Amer. Jour. of Math., vol. 78 (1956), pp. 761-790.

[A6] On the field of definition of a nonsingular birational transform of an algebraic surface, Annals of Math., vol. 65 (1957).

<u>S. Abhyankar and O. Zariski</u>

[AZ] Splitting of valuations in extensions of local domains, Proc. Nat. Acad. Sci. U.S.A., vol. 41 (1955), pp. 84-90.

<u>W. Krull</u>

[K1] Der allgemeine Diskriminantensatz, Math. Zeit., vol. 45 (1939), pp. 1-19.

[K2] Galoissche Theorie bewerteter Körper, Sitzungsberichte der Bayerischen Akademie der Wissenschaften, München (1930), pp. 225-238.

[K3] Allgemeine Bewertungstheorie, Crelle Jour., vol. 167 (1931), pp. 160-196.

[K4] Idealtheorie, New York (1948).

[K5] Ein Satz über primäre Integritätsbereiche, Math. Ann., vol. 103 (1930), pp. 540-565.

[K6] Zur Theorie der kommutativen Integritätsbereiche, Crelle Jour., vol. 192 (1954), pp. 230-252.

[K7] Die Verzweigungsgruppe in der Galoischen Theorie beliebiger arithmetischer Körper, Math. Ann., vol. 12 (1950), pp. 446-466.

D. G. Northcott

[N] Ideal theory, London (1953).

B. L. van der Waerden

[W1 and W2] Modern Algebra I and II, New York (1949).

O. Zariski

[Z1] Foundations of a general theory of birational correspondences, Trans. A. M. S., vol. 53 (1943), pp. 490-542.

[Z2] The concept of a simple point of an abstract algebraic variety, Trans. A. M. S., vol. 62 (1947), pp. 1-52.

[Z3] The reduction of singularities of an algebraic surface, Annals of Math., vol. 40 (1939), pp. 639-689.

[Z4] Local uniformization on algebraic varieties, Annals of Math., vol. 41 (1941), pp. 852-896.

[Z5] A simplified proof for the resolution of singularities of an algebraic surface, Annals of Math., vol. 43 (1942), pp. 583-593.

[Z6] Applicazioni geometriche della teoria delle valutazioni, Rend. Mat., Roma, Series I, vol. 13 (1954), pp. 1-37.

[Z7] Theory and applications of holomorphic functions on algebraic varieties over arbitrary ground fields, Memoirs of A. M. S., No. 5 (1951).

[Z8] On the purity of the branch locus of algebraic functions, Proc. N. A. S., vol. 44 (1958), pp. 791-796.

Lightning Source UK Ltd.
Milton Keynes UK
UKHW030145200522
403267UK00001B/47

9 780691 080239